ENCYCLOPEDIA OF
WHALES
DOLPHINS AND PORPOISES

ENCYCLOPEDIA OF
WHALES
DOLPHINS AND PORPOISES

ERICH HOYT

Principal photography by Brandon Cole
Illustrations by Uko Gorter

FIREFLY BOOKS

A Firefly Book

Published by Firefly Books Ltd. 2017

FIRST PRINTING

Publisher Cataloging-in-Publication Data (U.S.)

Names: Hoyt, Erich, author. | Cole, Brandon, photographer. Gorter, Uko, illustrator.

Title: Encyclopedia of Whales, Dolphins and Porpoises / Erich Hoyt ; photography by Brandon Cole.

Description: Richmond Hill, Ontario, Canada : Firefly Books, 2017. | Includes bibliographic references and indexes. | Summary: "In-depth profiles of 90 species of cetaceans, their life histories, biology and behaviors" – Provided by publisher.

Identifiers: ISBN 978-1-77085-941-8 (hardcover)

Subjects: LCSH: Cetacea. | Whales. | Dolphins.

Classification: LCC QL737.C4H698 |DDC 599.5 – dc23

Published in the United States by
Firefly Books (U.S.) Inc.
P.O. Box 1338, Ellicott Station
Buffalo, New York 14205

Published in Canada by
Firefly Books Ltd.
50 Staples Avenue, Unit 1
Richmond Hill, Ontario L4B 0A7

Library and Archives Canada Cataloguing in Publication

Hoyt, Erich, 1950–, author
Encyclopedia of whales, dolphins and porpoises / Erich Hoyt ; principal photography by Brandon Cole ; illustrations by Uko Gorter.

Includes bibliographical references and index.
ISBN 978-1-77085-941-8 (hardcover)

1. Marine mammals — Encyclopedias. I. Cole, Brandon, photographer II Gorter, Uko, 1962–, illustrator II. Title. III. Title: Whales, dolphins and porpoises.

QL713.2.H69 2017   599.503   C2017-901999-6

Front cover © Brandon Cole
Back cover, top left: © Tatiana Ivkovich/Far East Russia Orca Project (Whale and Dolphin Conservation); top right, bottom, spine and front flap © Brandon Cole
Author's photo © Alexander Burdin

Cover and interior design: Gareth Lind, LINDdesign

Printed in China

 We acknowledge the financial support of the Government of Canada.

Acknowledgments

For their superb work and collaboration, I would like to thank my editor Tracy C. Read and associate publisher Michael Worek, both of whom helped to shape the book, and designer Gareth Lind.

My gratitude goes to illustrator Uko Gorter, who let us feature his excellent cetacean illustrations, and to Brandon Cole, who provided principal photography. A huge thanks to contributing photographers Mike Bossley, the Far East Russia Orca Project (FEROP, WDC), Rob Lott, Edward Lyman, Rubaiyat Mansur, Flip Nicklin, Charlie Phillips, Robert Pitman, Alisa Schulman-Janiger, Mariano Sironi, Jean-Pierre Sylvestre and Deron Verbeck for sharing their work.

I am also grateful to Mark Simmonds, who read the entire text, and to Philip Clapham; both generously provided comments and corrections. Sections of text were pored over by Regina Asmutis-Silvia, Mike Bossley, Philippa Brakes, Jim Darling, Sarah Dolman, Olga Filatova, Astrid Fuchs, Nicola Hodgins, Miguel Iñíguez, Charlie Phillips, Donna Sandstrom, Hans Thewissen, Vanesa Tossenberger, Courtney Vail, Colleen Weiler, Hal Whitehead and Vanessa Williams-Grey.

I would also like to thank indexer Angus Barclay and colleagues and friends Tundi Agardy, Jim Borrowman, Alexander Burdin, Chris Butler-Stroud, Niki Entrup, Ivan Fedutin, Lesley Frampton, Sylvia Frey, Tatiana Ivkovich, Dan Laffoley, Rob Lott, David Mattila, Naomi McIntosh, Giuseppe Notarbartolo di Sciara, Kotoe Sasamori, Michael Tetley and Olga Titova.

This book would not have been possible without timely, amazing help from my angels, Jade Costin-Barrett, Sam Paull, Abby Smith, Annie McGarvey, Linda Denslow and Paul Millfield, as well as my family, Moses, Magdalen, Jasmine, Max and Sarah, and my sisters, Chrissie, Victoria, Karres and Sissy.

This book is dedicated to Sarah Elizabeth Wedden, with all my love.

Contents

Previous spread: A herd of swift-moving long-beaked common dolphins travel together in the eastern North Pacific Ocean off California.

Foreword

~~~~~~

Back on the sea: The wind on my face, the smell of salt water, the crack of an explosive spout off the bow, the rapid-fire clicking of cameras as the broad back and dorsal fin appear, the plop of a hydrophone entering the water to eavesdrop on a trumpeting whale song and the local dolphin dialect. And then, suddenly, the ocean beneath us comes alive with rumbles, clicks, squeals and screams.

We have entered the world of whales, dolphins and porpoises—the so-called cetaceans.

For the past 40 years, I have been lucky to experience this moment many times over, yet I never tire of it. Each time brings with it a frisson of excitement—the chance of a new finding or the discovery of an unusual behavior in the company of fellow researchers and whale friends.

Living at the edge of what we understand about cetaceans has been a wild ride. With little information to go on, humans initially considered these creatures as other-worldly beings. After generations of study, we began to view them as social mammals and predators that had evolved to live in the sea. But in recent years, we have glimpsed just how unusual they are in their own right. The richness of detail that we're uncovering as we explore their lives in the wild is at the heart of this book.

Over the course of my time on the ocean, there has been a revolution in our thinking and understanding about wild cetaceans. In the late 1960s, the "Save the Whales" movement put whales in the public eye. The advent of whale watching, a recreational activity that first became popular in California but has migrated to all corners of the world, brought whales out of the "sea monster" era and transformed the ocean into a much friendlier place, alive with possibilities.

Scientific breakthroughs typically come from specialists working at the narrowest end of their specialty. By contrast, the breakthroughs in our understanding about whales are the result of work done not only by specialists but also by generalists and members of the public. While what we've learned about whales is largely a result of long years of dedicated fieldwork, it is also a triumph of the growing citizen science movement.

Many scientific papers on cetaceans have added to our knowledge in recent years. While this book draws on those papers, it aims primarily to take you out into the field to observe, learn about, participate in and enjoy the world of wild cetacean research. At the same time, once there, it will be hard not to consider the hazards cetaceans face every day in the wild, as well as to ponder the ultimate fate of the ocean. Whether humans can change those outcomes remains to be seen.

—Erich Hoyt
Bridport, Dorset, England

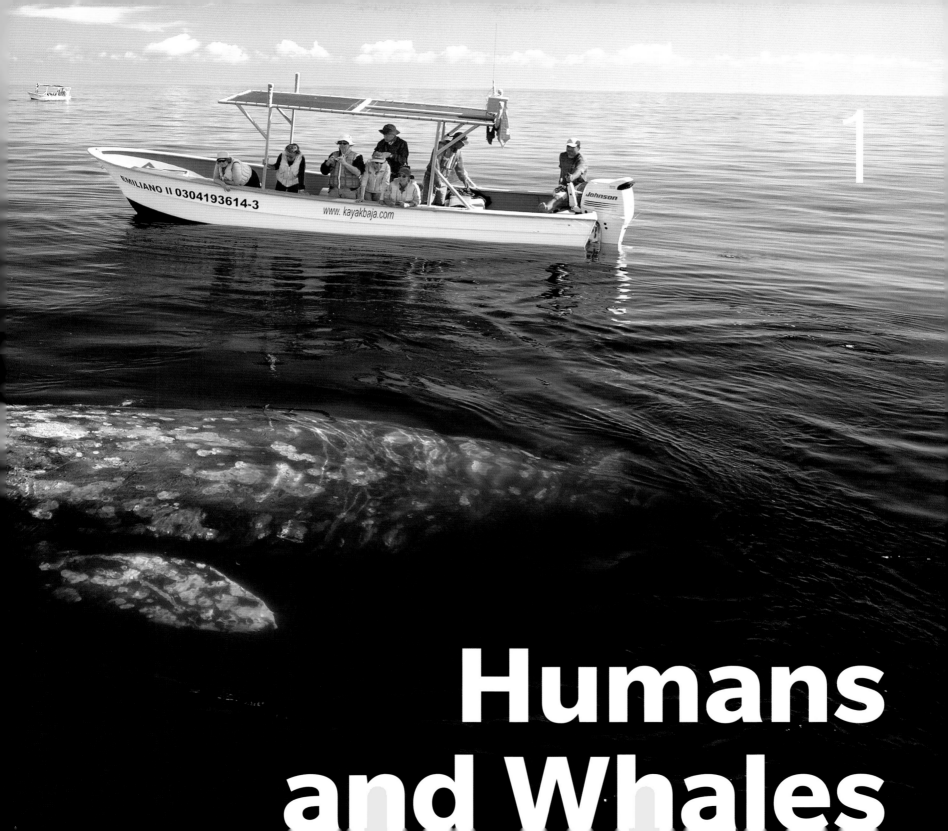

1

Humans and Whales

HUMANS AND WHALES
Our Shared History

Whales, dolphins and porpoises are air-breathing social mammals that nurse their young with milk and exhibit intense parental affection. Dolphins have befriended human children. Fishermen can tell dolphins apart from nicks in the tails, and this allows individual dolphins to be aged to 25 to 30 years or more.

— From Aristotle's *Historia Animalium*
(paraphrased from the original Greek, c. 350 BC)

The first marine biologist, the first naturalist guide and the first philosophical ruminator, Aristotle was intent on classifying known life on Earth and in the water, and he made a good start. His fundamental writings about the natural world show that he was a skilled gatherer and sifter of information. He interviewed one and all, took notes and then applied good horse sense. He got out into the field to explore nature, to discover truth and to make firsthand observations to enrich what he heard from others.

Given what we know about whale and dolphin societies today, not much of Aristotle's brief summary of whale and dolphin social life and insight into individual identification surprises us. What *is* astonishing is that before Aristotle and for nearly 2,000 years after him, no one came close to making

Previous spread: A gray whale surfaces to breathe near a whale-watching boat in a Baja California lagoon in El Vizcaíno Biosphere Reserve, Mexico. Researchers from the Far East Russia Orca Project, facing page, photograph killer whales in Avacha Gulf, Kamchatka, Russia.

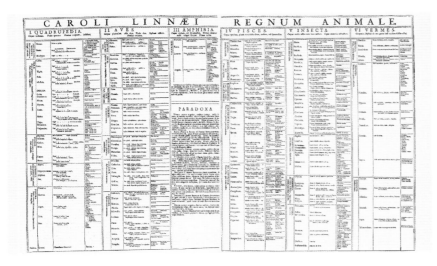

Linnaeus's classification of the animal kingdom, above, initially ignored Aristotle's research on marine mammals. Typical scrimshaw carving on a sperm whale tooth, top right, shows a sail-powered whaling ship. In earlier centuries, whale oil in various grades, bottom right, was used to light lamps, heat homes and lubricate watches.

so many new findings about animals or sensibly disputing his observations. Even the distinguished father of taxonomy Linnaeus (Carl von Linné) appears to have side-stepped Aristotle's marine research, classifying whales and dolphins as fishes in the first nine editions of his *Systema Naturae* in 1735. It wasn't until the 10th edition of the book, published in 1758, that whales were put into the mammal class.

Yet along the timeline that runs from Aristotle to Linnaeus, the human relationship with whales had become defined as chase and kill. Whales were eaten, and they were turned into oil to light lamps, heat homes and lubricate machinery.

Baleen—the flexible keratinous substance found growing in the mouths of some whales—was trimmed and shaped to manufacture umbrellas and corsets. Sperm whale teeth were used in detailed carvings known as scrimshaw. A wide range of other products emerged that varied from country to country and century to century. The bottom line was that until the mid-20th century,

A typical factory where whale blubber was rendered into oil: The blood is gone, but the massive machinery of death remains.

there was serious money to be made in whaling.

Commercial whaling started with the Basques around 1000 AD. They worked their way through the North Atlantic right whale population, which to this day has not recovered and hovers around 500 individuals in the entire North Atlantic. By 1600, western European whalers had moved on to bowhead whales in the Arctic. In the late 1600s, whaling started moving west. Yankee whalers took over, focusing on North Atlantic right and humpback whales and then sperm whales in the early 1700s. After they depleted the whale populations along the U.S. east coast, they began moving farther afield to Greenland in the north and to the

Norwegian Sven Foyn invented the exploding harpoon in 1864—a technological innovation that, with subsequent additions, would seal the fate of close to 3 million whales across the world ocean.

Caribbean and open Atlantic in the South, and then to Hawaii, Japan and the rest of the Pacific.

Two developments sealed the fate of the whales. The first was Norwegian Sven Foyn's 1864 invention and 1870 patenting of the explosive harpoon, a tool of destruction whose use quickly brought populations of the slower whales to the brink of extinction. Then, in the first decades of the 1900s, the development of fast catcher boats and factory ships opened up the Antarctic and enabled whalers to catch the speedy blue, fin and sei whales. In terms of numbers of whales killed

in just a few decades, the mid-20th century was the worst chapter. One by one, species were driven to commercial extinction—the point at which the revenue from the number of whales killed is less than the cost of finding and catching them.

Many whalers gave up at this stage, while the early International Whaling Commission struggled and failed to protect the remaining populations. Pirate whaling by non-IWC member states persisted, but it was dwarfed by illegal whaling from IWC member states themselves. Illegal whaling by Russian Soviet fleets resulted in the slaughter of tens of thousands of whales, in a blatant violation of rules regarding protected species and undersized animals. These secret catches included protected blue, North Pacific right, bowhead, humpback and sperm whales. The illegal Soviet whaling campaign was revealed only some years after the breakup of the USSR. In the 20th century alone, an estimated 2.9 million whales were killed by all nations combined.

But times were gradually changing. In the late 1940s, Carl Hubbs, a widely published freshwater fish researcher from Michigan, arrived at the Scripps Institution of Oceanography, in San Diego. He quickly expanded his studies to the marine world, counting gray whales as they migrated up and down the California coast. Indeed, Hubbs introduced a generation of students to the joys of whale watching. By the 1950s, land-based whale watching by the public from headlands and lookouts along the California coast was growing. In 1955, the first whale-watching boat hung out a sign at the wharf in San Diego, offering trips at the cost of $1 USD "to see the whales—real ones."

Hubbs and his students were ahead of their time. Meanwhile, wholesale slaughter was taking place in the Antarctic, where the blue, fin, sei, humpback and southern right whale species were being nearly eliminated from the vast Southern Ocean. Along the Pacific coast of North America, too, as late as the 1960s, hundreds of whales were still being killed every year. Only in 1971 did the last California whaling station owned by the Del Monte Fishing Company based at Richmond, on San Francisco Bay, finally shut down.

By the late 1960s, "Save the Whales" banners were starting to appear. Whales became an icon of the environmental movement in the United States and Europe. The implied message: If we couldn't save the whales, how can we save ourselves? Greenpeace was formed to oppose nuclear testing in Alaska and the South Pacific and soon recognized the power of the whale as a symbol. Some of the original members, such as Robert Hunter and Paul Spong, became interested in wild whales, though it wouldn't be until 1975 that Rex Weyler of Greenpeace took the unforgettable photograph of their team in a small boat facing up to the Soviet harpoon guns in the open North Pacific.

In summer 1975, the Russian whaling fleet was busy harpooning sperm whales in the North Pacific. In an effort to stop them, Greenpeace drove their inflatable boat between the harpoons and the whales. The whales died, but the photos went viral, fueling the "Save the Whales" movement.

Even as the whale was becoming the symbol of the environmental movement, however, we knew almost nothing about how whales lived in the wild. Anatomical studies from the scientists who recovered whale carcasses or spent time on whaling boats told us that they had big brains. The U.S. Office of Naval Research had learned about dolphin sonar beginning in the early 1950s, with Winthrop Kellogg's research at Florida State University. Before that, in 1938, the first aquarium to train dolphins to perform tricks opened up at

A spinner dolphin off Hawaii suffocates in a piece of discarded fishing net. Nets have killed millions of cetaceans and unwanted fish in the process of catching target fish species. But broken pieces of net can also be lethal.

of pole and line fishing, the newer nylon nets began killing both tuna and dolphins in huge numbers. In total, the tuna fishing industry was responsible for more than 6 million dolphin deaths—more than double the whaling toll in the 20th century. Today, the dolphins killed by tuna nets total about 1,000 to 3,000 per year. Buying only "dolphin-safe tuna" has greatly reduced the number of deaths, but the number killed is still too high, and even temporary captures in nets may result in stress and social disruption to the dolphins that are released alive.

The political response to this heightened awareness of cetaceans and to the impact of the rapacious whaling and fishing industries was not instant, but the United States soon took the lead in pioneering two landmark pieces of legislation that reverberate strongly to this day: the Marine Mammal Protection Act (1972) and the Endangered Species Act (1973).

Around the same time, exciting new findings about whales and dolphins started to emerge. Researchers Roger Payne and Scott McVay discovered the songs of the humpback whale. Along with their paper in *Science* in 1971, they released an album of songs that became a bestseller. Meanwhile, Payne and his then wife Katy

Marine Studios in St. Augustine, Florida. In the 1960s, the number of dolphin aquaria grew rapidly and, besides the common bottlenose dolphin, killer whales were introduced for the first time to an impressionable public. The public interpreted the ability of dolphins to learn to do tricks as a sign of intelligence.

At the same time, dolphins were being targeted by tuna fishermen who set their seine nets around big dolphin schools, especially in the Eastern Tropical Pacific, to catch the large yellowfin tuna that swam below the dolphins. After years

moved with their family to Patagonia, in southern Argentina, to immerse themselves in the world of southern right whales, which were then beginning their long climb back from near extinction.

Observing the whales every day from the cliffs, the Paynes began to see that the black and white

In the early 1970s, researchers working independently in the United States, Canada and Argentina began to take and catalog photographs to distinguish individual (clockwise from top left) humpback, gray, southern right and killer whales. This was the start of a revolution in our understanding of the lives of whales and dolphins.

markings on their heads were individually distinctive and could be used to tell animals apart. It wouldn't be until November 1975, at the seminal National Whale Symposium, in Bloomington, Indiana, that researchers working in different parts of the world realized that they had simultaneously cracked the code of how to identify individual whales and dolphins. They found that they could use individual differences in the dorsal fin (killer whales and various dolphin species), the pattern on the tail flukes (humpback whales), the shapes and arrangement of the callosity patches on the head (right whales) and pigmentation patterns or mottling on the broad back (gray and, later, blue, fin, and minke whales).

More than anything, photo identification (photo ID) was the tool that launched the modern study of whales, dolphins and porpoises in their natural habitat. In 1973, I was a young filmmaker and prospective environmental journalist spending my first summer living near killer whales off northern Vancouver Island and getting to know whales with names like Stubbs, Nicola, Hooker and Tsitika by their appearance and habits. Canadian marine mammal scientist Michael Bigg had found that sharp photographs were enough to identify any individual orca, and over the next decades, he and his colleagues created a catalog to show the killer whale matrilines, pods, clans and communities. These maps of the social relationships would be created with other whale and dolphin species too.

In 1979, researcher Jim Darling matched his photos of Hawaiian humpback whale tails to several of Chuck Jurasz's photos of tails from Southeast Alaska, confirming a migratory link. Around the same time, working in the Gulf of Maine, Steven Katona and his colleagues began matching humpback photos to the Caribbean. Later, Darling matched his Hawaiian humpback tails with Mexican tails, thus proving that the humpbacks were changing breeding grounds, traveling some years to Hawaii and others to Mexico. Darling was also surprised to learn from Katy Payne that all the humpbacks in the North Pacific were singing essentially the same song, although, over the season, the song could be heard to be slowly changing, evolving into a new song for the next breeding season.

The ability to recognize individual whales on sight and through photography opened the door not only to studies of migration patterns, abundance and distribution but to determining their ages and understanding their social behavior. These findings eventually led to insights into what researchers began to call "whale culture."

I remember standing on a rock face overlooking Johnstone Strait on northern Vancouver Island that first year in 1973 with members of Stubbs's pod all around, eagles slicing the sky and snatching salmon driven to the surface by the orcas. The orcas visited us four or five times a day, and we felt part of this developing adventure. We

watched the old matriarchs Stubbs and Nicola take care of the calves. Later, we watched them come into the rubbing beach and take turns rolling and rubbing on the smooth bottom. We didn't call it culture then, but as we learned about orcas in other parts of the world, we realized this behavior was unique. Only this northern community participated in rubbing. It was a piece of culture they shared. Later, I would see orcas training their young to beach themselves at Punta Norte, Argentina, in preparation for the sea lion pupping season.

Killer whales take turns rubbing against smooth pebbles and rolling on the sand at a rubbing beach on northeastern Vancouver Island—a cultural behavior practiced only by the local killer whale pods of the northern fish-eating community.

Since then, I have returned nearly every year to some whale spot or another in more than 50 countries to eavesdrop on these wild creatures and to learn more about them. Everything—and yet nothing—surprises me about them. We still have a long way to go before we achieve a deep understanding. But we are learning, day by day, and enjoying every minute of this wild ride.

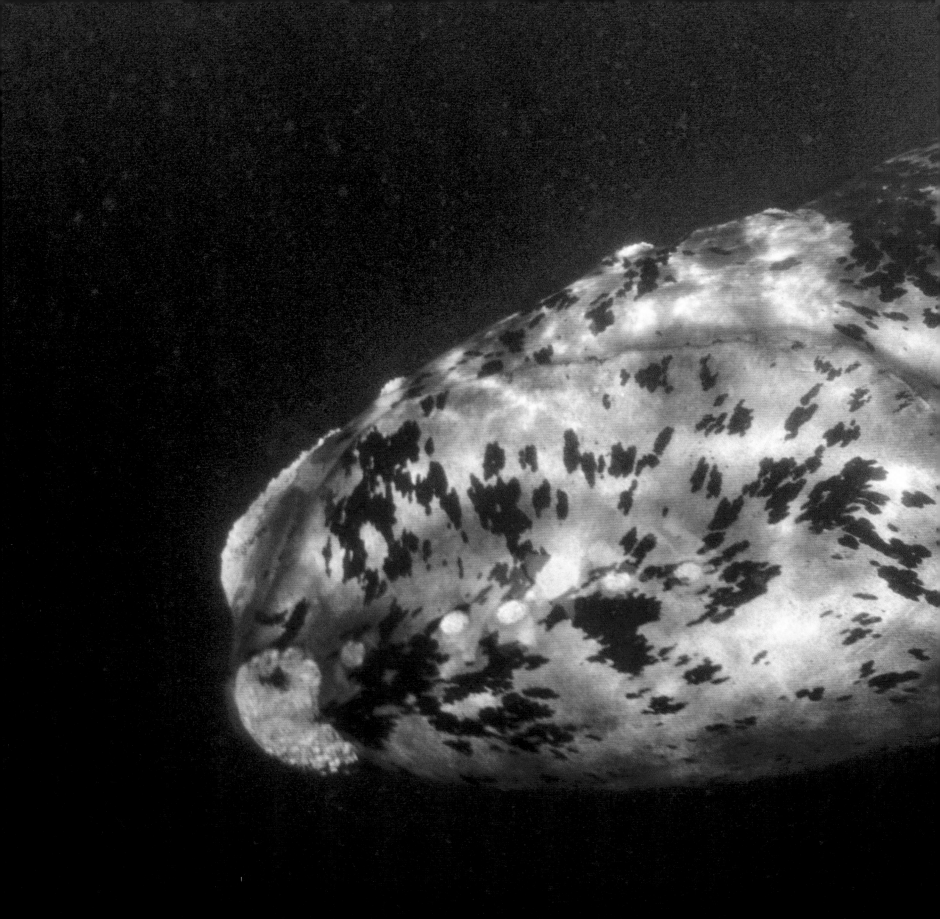

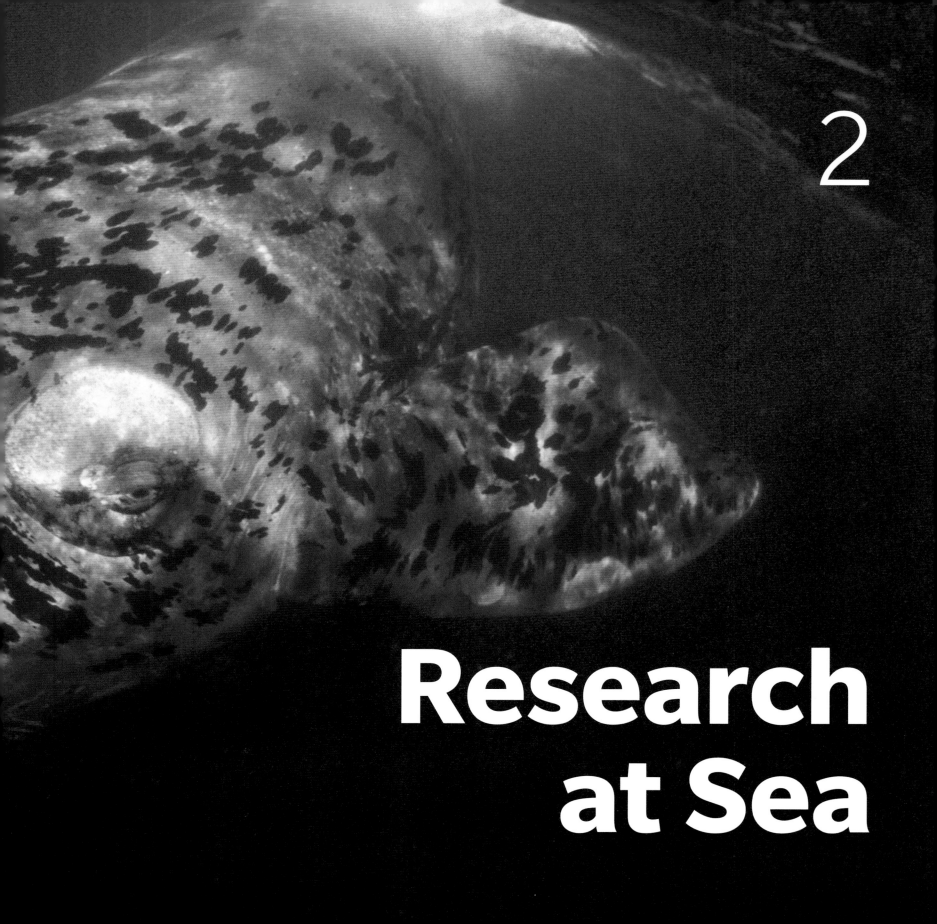

2

Research at Sea

Studying Whales, Dolphins and Porpoises

For many years, researchers and whale-watching skippers and naturalist guides have met up at an annual spring "naming of the whales" party on Cape Cod. As photographs of the underside of humpback whale flukes are projected on a screen, participants propose new whale names based on the intricate black and white fluke patterns on each individual. A vote is held, and the winning name is then toasted—and a new whale is added to the catalog. That individual's life fortunes can then be followed every time the whale reappears. Special interest is given to females as their potential calves can be followed as well.

Photo identification

In the early 1970s, researchers working independently on southern right whales in Argentina, gray whales and killer whales in British Columbia and humpback whales in the North Atlantic began photographing individuals and noticing marks that would allow individual identification. The method was called photographic identification—photo ID. In time, some whales could be identified on sight, but most required the careful consideration of photographs that were cataloged according to date, time, associations and behavior. In fact, these whale researchers had borrowed a technique used

Previous spread: A young white southern right whale in the waters off Patagonia, Argentina, shows typical curiosity. Facing page: Researchers love to see a humpback whale breach clear of the water, but they wait for the deep dive when a humpback shows its flukes. The underside of the tail flukes is unique to each whale.

individuals. Humpback whale catalogs, by contrast, cover ocean basins. The largest, the North Pacific humpback whale catalog, includes more than 18,000 photo-ID records of nearly 8,000 individual humpback whales.

The underside of the tail flukes is the key to photo identification with humpback whales. Here, the pigmentation ranges from nearly all white to nearly all black, with spots, splotches and lines in black or white that create unique patterns. But different methods have been developed for other species of whales and dolphins. With killer whales, the identification emerges mainly from the nicks and scratches present on the trailing edge of the dorsal fin, as well as from the shapes of the dorsal fin and saddle patch; even the eye patch can aid identification. Researchers identify bottlenose and other dolphins from the markings on the trailing edge of the dorsal fin and

Members of the author's research team struggle to take photo IDs of killer whales off Kamchatka when the orcas approach too close to the boat. The team now has a catalog of more than 2,000 individual orcas, many of them seen year after year.

decades earlier with giraffes, elephants, lions and chimpanzees in Africa. It had taken longer to adapt the technique for animals at sea that were often far offshore; whales move fast and often spend as little as 5 percent of their time at the surface. But researchers soon realized that individual identification could unlock social behavior, migration and many other details of cetacean biology.

Today, there are photo-ID catalogs for scores of populations of whales, dolphins and porpoises. Bottlenose dolphin and killer whale catalogs may contain a few hundred or as few as 25 to 30

its overall shape as well as any other body marks. Sperm whales can be individually identified mainly from markings on the flukes and occasionally from body scars.

The three species of right whales—the North Atlantic, North Pacific and southern right whales—lack a dorsal fin and are instead

The extensive white coloring on this mother and her calf, facing page, marks them as Southern Hemisphere humpbacks. Individual identification, however, relies on the researcher photographing the underside of the tail flukes.

Identifying Humpbacks

Humpback whales can be individually identified via a sharp photograph of the underside of their flukes, which they lift before diving after a sequence of several breaths. Researchers from the Russian Cetacean Habitat Project of the Whale and Dolphin Conservation group took these photo IDs off the Russian Far East, mainly in the Commander Islands in the far northern North Pacific. They have given names and numbers to the whales, often based on the appearance or inspiration that comes from the black and white fluke patterns and the shape of the flukes. From left to right in the top row are Mervent (1), Sandpiper (2) and Glacier (3). In the middle row are Tuna (4), Chained (5), Blind (6) and The Shadow of Hamlet's Father (7). In the bottom row are In Lace (8), Hares (9), Jonathan [Livingston] (10) and The Fisher King (11). The female In Lace has been sighted every summer since 2009, and in 2016, she appeared with a calf. Work is underway to match these photos to photos taken on breeding grounds, which will provide information about their travels and possible breeding groups.

The male killer whale Taku, K1, center, is distinguished by the two notches on the trailing edge of his dorsal fin. These notches were made by researcher Michael Bigg in 1974 when Bigg was first learning to identify killer whales and wanted to verify that markings on the trailing edge were permanent. After a brief period in the capture nets off southern Vancouver Island, Taku was released and returned to K pod, part of the southern resident community.

markings but not enough for purposes of identification. Instead, individual identification is made possible by the subtle patterns showing swatches of pigment (especially in minke, sei and fin whales) or mottled pigment (in blue whales). Researchers have shown that these pigmentation patterns change little from year to year, which means it is possible to recognize the whales when they appear again.

Scars are also distinguishing features, as with the cookie-cutter shark scars seen on the flanks of sei whales. And no matter the species, some individuals carry marks from other injuries that also prove valuable for photo ID. These include propeller marks on the back, flukes or fins; net marks or other fishing gear injuries on the back; and killer whale bite marks on flukes. One female orca from the northern fish-eating community off Vancouver Island had a grapefruit-sized tumor on the side of her head. She was observed over several seasons with this tumor; even so, her unique dorsal fin was more visible.

There are established conventions for photo IDs. With many species, photos are usually taken of the whale's left side, unless both sides can be obtained. The date and place are always noted, as is the name of the photographer. In the early days, photo-ID catalogs were typically printed, and some have since been reproduced on laminated sheets for easy access in the field. Most catalogs, however, are now online and can be accessed by phone, tablet or laptop.

identified from photographs that show individual patterns of callosities—roughened patches of dark skin inhabited by visible colonies of white or red cyamid parasites. The callosities are arranged at various spots on the head, especially on the chin, lips and near the upper jaw on top of the head. The latter callosity is called the "bonnet."

Identifying individuals among the more streamlined baleen whales—the minke, sei, fin and blue—requires sharp, well-lit photos that show the broad expanse of the back, anchored by the dorsal fin. The dorsal fin itself sometimes has a few

These catalogs contain photographs taken by a wide spectrum of people. As whale-watching tours have spread around the world, operators have invited researchers on board and have encouraged their guides to take photographs. Whale-watching skippers and guides spend more time on the water than many researchers and often get some of the best photos. Tour passengers also take photographs, and as a result, most catalogs have valuable photos contributed by members of the public.

Counting is the starting point for researchers who need to advise wildlife managers and government ministries about abundance and distribution in a local coastal area. The first studies of wild whales made use of photo ID largely as part of a counting exercise that would help establish whether a species was endangered or vulnerable. Nonetheless, it might take years to get a population estimate.

Eventually, photo ID grew into a way to understand how whales and dolphins live. To study their lives, researchers have to determine whether they are males or females, sexually mature or post-reproductive, how they are related to each other and whether the same animals are seen during a field season and from year to year. Researchers need to find out what whales eat and what their essential needs are for habitat. The photo-ID methodology became the foundation for much of the above and more. For instance, the identification of vocalizations in dialects particular to each killer whale pod would never have happened without individual photo identifications.

Transect studies

Another whale-counting technique is known as the transect study. Because this study is conducted from a steadily moving boat or from the air, it can cover a much larger area and provide a faster result in terms of estimated numbers. It can also be expensive. The study randomly chooses transect lines, following a survey design that will give a good estimate of the abundance of each species. Experienced whale researchers take turns using binoculars to watch fixed portions of the sea and to identify and count all the individuals of each species they see, which often includes other marine mammals, seabirds and turtles. Since waves and whitecaps limit the visibility and diminish the number of sightings, these surveys can only be conducted in fairly calm conditions. The results provide a snapshot of the abundance of multiple species on a given day and in a given season and so need to be repeated in other seasons and again every few years to track abundance trends.

Transect studies and photo ID are both valuable tools for researchers, but because photo ID has other uses in addition to census making, it has become much more of a multi-purpose tool. As such, it is used to monitor births and deaths within a season and over multiple years at a fine scale. Associations between individuals can be logged and studied for many years. Individual whales identified in locations at different times, sometimes thousands of miles apart, help to document the long migrations some whales make and the interconnections between different areas. For example, thanks to photo ID, researchers have discovered examples of individual southern right whales and humpback whales changing breeding grounds in alternate years.

Acoustic research

Acoustic studies are another way to monitor whale and dolphin populations. Underwater microphones known as hydrophones pick up sounds that echo for miles through the ocean's liquid universe. They are connected by a waterproof cable to a small digital recorder or even a smart phone on board a boat or on land. These devices detect a wide variety of whale and dolphin sounds. They are also used by researchers and whale-watching operators to help locate whales and dolphins long before we're able to see them. Some hydrophones detect whale sounds that are audible to humans, but some dolphin and porpoise sounds extend well above our hearing range, while blue and fin whale sounds register below it.

Some hydrophones are portable and can be easily dropped over the side of the boat once whales or dolphins are spotted. Alternatively, they can be towed behind a boat in what's called a hydrophone array (three or more hydrophones), which allows researchers to pinpoint the location of the sound. Typical use requires a boat that is either stationary or traveling slowly, but hydrophones can also be moored or fixed on the seabed as a passive acoustic monitoring system.

These passive acoustic systems are connected to recorders and speakers on land. A network of hydrophones is monitored in real time in the Johnstone Strait-Blackfish Sound area of northern Vancouver Island, Canada. As pods of killer whales pass through, the sounds are picked up at OrcaLab on Hanson Island and other locations. Researchers are thus able to monitor known killer whale pods, each with its own distinctive dialect, as they travel through the region, even at night and during winter storms.

In October 2015, researchers Paul and Helena Spong heard familiar northern community killer whale sounds on the OrcaLab hydrophones as well as sounds that they had never heard before off northern Vancouver Island. The new sounds, they soon realized, came from southern community killer whales that had never been known to visit the area. The two communities—separate populations, or breeding units—did not interact, but hearing their sounds together on the hydrophone was a rare experience.

Other passive acoustic systems are designed to remain in one location. Sonobuoys are deployed for a few hours to pick up and broadcast sounds by radio to a ship. A newer tool, called the C-POD (developed from an earlier generation tool called the T-POD) is a passive acoustic system that can be left at sea for months at a time. The C-POD consists of an omnidirectional hydrophone, a digital processor, a battery pack and analysis software in a watertight package.

A C-POD can detect the biosonar, or echolocation clicks, of toothed whale species such as sperm whales, belugas and narwhals, and dolphins and porpoises. Individual species are identified by the type of click. Harbor porpoise clicks, for example, can be detected at a range of approximately 1,000 feet (300 m), but 330 feet (100 m) is the usual limit for species detection. The device can provide time data on cetacean activity to show presence or habitat usage in an area. This data is then used to determine distribution, density, population trends and echolocation behavior over time. Sometimes, feeding activity can be determined. From time to time (usually no longer than every four months), researchers retrieve the C-POD and download the data stored on a memory card.

C-PODs are proving valuable in environmental impact assessments (EIAs) and the collection of baseline data prior to, during and after industrial activities related to fisheries interactions and the installation of drilling rigs and marine renewable energy devices.

Some acoustic research has been conducted at an ocean-wide scale. During the Cold War, the U.S. Navy used an extensive network of stationary hydrophones on the seafloor to track Soviet submarines across the North Atlantic and in the North Pacific. After the collapse of the USSR in 1991, the U.S. Navy allowed researcher Christopher Clark, director of the Bioacoustics Research Program at Cornell University, to listen for whales on the Sound Surveillance System (SOSUS).

Clark was amazed when SOSUS allowed him to pinpoint the location of blues, fins, humpbacks and minke whales on maps. He found that their sounds could be heard over thousands of miles of ocean. He could "see" whales swimming toward a seamount some 300 miles (500 km) away and then changing course and heading to a new oceanographic feature. It seemed that whales had maps of the ocean in their heads, formed by their enhanced acoustic abilities and replenished by what Clark thinks must be "acoustic memories," akin to our visual memories. Yet the blues, in particular, were working on a grand scale of time and space barely comprehensible to humans. To a blue whale, a neighborhood might effectively stretch over thousands of miles of ocean.

When a blue whale starts singing, bellowing its loud, low notes, one note—traveling 4.4 times faster and much farther than sound in air—will, in two minutes, carry 100 miles (160 km) through the ocean. Clark tracked the sounds of a blue whale

Far East Russia Orca Project researchers drop their hydrophone into the ocean to listen for killer whales off Kamchatka.

off Newfoundland being picked up some minutes later on the hydrophone off Bermuda. After an hour, the sound could theoretically be picked up 3,000 miles (4,825 km) across the ocean by a hydrophone or by another blue whale.

But Clark also tuned in to a lot of noise, and he expressed concern that the calls of blue whales might not be getting through to other blue whales. The ocean is indeed getting noisier due to ship traffic, hydrocarbon exploration and navy sonar. If female blues can't hear males singing through

OrcaLab on Hanson Island off northeast Vancouver Island is linked to underwater hydrophones and has a lookout station to monitor and record passing killer whales.

and other countries. Researchers from the Irish Whale and Dolphin Group and the U.K.'s Sea Watch Foundation and Whale and Dolphin Conservation monitor mainly nearshore bottlenose and other dolphins and porpoises from land-based lookouts around the coast.

Some research teams depend on land-based observers who are situated on high vantage points to alert them before they proceed with boat-based work, thereby saving time and money. The tools of the trade include high-power binoculars, monoscopes and theodolites for determining positions, all mounted on tripods.

The first scientists who studied the gray whale started by counting individuals from shore in the late 1940s. As the endangered grays migrated past southern California every year, and the numbers of both whale counters and whales climbed steadily, the news seemed to be good. Counting whales from land, however, provides only minimum estimates. If it is possible to calculate the percentage of whales that can be seen from land, in addition to those that either don't migrate or are traveling far out to sea, then it is possible to make estimates of the total population. To learn more, researchers eventually took to boats to study gray whales on their feeding grounds around Alaska as well as in the breeding lagoons in Baja California, Mexico.

Because it takes place out of sight and too far away to influence the whales, land-based research provides great opportunities for observations about behavior.

the noise in their habitats, then they may be losing breeding opportunities. And if their communication range becomes limited, it also means their habitat is being systematically reduced.

Land-based studies

Counting whales from land rather than from boats and planes is the least expensive way to conduct a census. Land-based observations are useful for the few species that migrate close to land, such as gray whales off the west coast of North America, humpback whales along the east coast of Australia or spinner dolphins that rest inshore in the island archipelagos and coastal areas off Brazil, Hawaii, southern Japan, Australia, Egypt

Science for Conservation

Much of scientific enquiry is hypothesis-driven: A scientist comes up with a theory and then tries to prove or disprove it. But many field biologists conduct their science by amassing large amounts of data, usually over a period of years, and then looking for the patterns. This is how biologists going back to Aristotle have worked. Darwin spent years "doing his barnacles" before coming up with his ideas on evolution by the mechanism of natural selection. His five-year voyage as a young man on the *Beagle* around the tip of South America and up to the Galápagos allowed him to amass substantial evidence.

Darwin did not have the benefit of the genetics revolution, but his work set the stage for it. Still, it has taken many decades for genetics work to be incorporated into field studies in a cost-effective manner.

The genetic analysis of whale tissues has provided wide-ranging determinations and has had substantial conservation value. Combined with photo ID and other techniques, it can provide new insights into species, including relationships within populations and populations within species. It is possible to define new species, which become the foundation for evaluating the conservation status that leads, in turn, to a management response.

Genetic studies are sometimes combined with analysis of blubber for contaminant levels. As might be expected, killer whales that feed on marine mammals carry higher levels than fish-eating orcas. Belugas in the St. Lawrence carry some of the highest concentrations of PCBs and other contaminants—a cocktail of chemicals from all the upstream industries produced in cities on the shores of the Great Lakes, such as Chicago, Detroit, Buffalo, Toronto and Montreal. Skin samples can also be analyzed for stable carbon and nitrogen isotope ratios. Stable isotope ratios help reveal the diet and trophic level of whales—where they are eating on the food pyramid. Such studies also provide insights into population structure and migration routes and can be used with other information to confirm findings about foraging ecology.

Genetic and stable isotope studies can sometimes be undertaken with tissue found floating on the surface, for example, after a humpback whale jumps out of the water, and even from carcasses that wash up on a beach. Researchers have purchased whale meat in the markets in Japan and Korea to obtain DNA samples—these have been used to identify Endangered or other species that may have been caught illegally or those that have been passed off as another species. Researcher C. Scott Baker began this work with whales in Japan and Korea in the early 1990s, and since then, DNA reality checking has become more and more useful for monitoring the trade in many kinds of endangered wildlife products.

The most common way to gather genetic samples is through biopsy darting of known individuals as part of a larger photo-ID study. Approaching the whale at close range in a boat, the researcher uses a crossbow or what looks like a rifle to fire a dart with a short steel tip into the flank of the whale. The dart bounces off the whale and floats in the water until it is retrieved.

Another technique for monitoring swimming

Diet Detection

Examining the stomachs of stranded whales and dolphins was the first method researchers used to get clues about the diets of cetaceans. In 1861, Danish researcher Daniel Eschricht opened up the stomach of a 21-foot-long (6.4 m) mature male killer whale washed up on the Denmark coast and found pieces of 13 porpoises and 14 seals. Eschricht subsequently produced a paper that would be cited hundreds of times and is still a common reference.

Today, if a whale strands on a beach, a researcher needs to reach the carcass as soon as possible in order to examine the stomach and teeth and take tissue samples. What resides in a dead whale's stomach suggests that a given item was in the diet just before the individual died. Specialists are needed to confirm the prey species, which are sometimes determined from the otoliths, also known as "fish ear bones," as well as from squid beaks. However, since the remains are confined to the hard parts of prey, other items that are too far digested as well as soft-bodied prey are missed completely.

Occasionally, a researcher may witness hunting behavior or see a whale or dolphin catching and eating a fish at the surface. Photographs or video can help to verify an account, but determining the exact species for smaller prey is problematic. Exceptions are killer whales that have been photographed devouring known sea lion species in Patagonia or taking down a blue whale. In one case, photographed from the air, several killer whales lunged and bit into a fast-swimming blue whale, removing large pieces of its back.

Researchers can also observe and analyze diving patterns of whales or dolphins in a particular area if they know the potential prey that live at those depths. For example, many beaked whales are found in areas with depths corresponding to the presence of squid, and the length of their dives tends to confirm that the whales are diving deep.

Some North Pacific researchers watch for shimmering salmon scales that float on the surface in the midst of killer whale feeding bouts. Scooping up the scales with a net, they preserve them in alcohol for lab analysis to determine the precise salmon species. These "crumbs"— small or partly eaten remnants of a prey item—are then subjected to DNA or other analysis.

Another new assessment method, sometimes combined with others, is stable isotope analysis of tissues taken from a whale with a biopsy dart, which allows a researcher to estimate the diet and trophic level of the prey. This method has been used to determine whether humpback whales, in a given area, generally rely on krill or whether they are feeding higher up the food chain. Compared with stomach studies that give a snapshot of the last meal, isotope analysis enables insight into the whale's prevailing diet over its last weeks or months.

The diet of the deep-diving ecotype of killer whales called "offshores" that live off British Columbia in the North Pacific was a mystery until Canadian researchers pieced together clues. Stomach contents, fatty acid and isotope analysis pointed toward a diet that included halibut and sharks. An examination of the mouth of a stranded, dead offshore orca revealed teeth that were more worn down than in other killer whale ecotypes, indicating the orcas' preference for shark. Sometimes, the researchers noticed oily patches of water on the surface above the areas where the killer whales were diving deep. Such oil slicks might form, researchers speculated, if the whales down below

Some ecotypes of killer whales in the Antarctic specialize in hunting the Weddell seal, above. When researchers witness orcas catching seals, whales or other prey at the surface, they can document their hunting methods and diet preferences in detail.

were ripping apart sharks to get at their large livers.

All the above methods have a place in researchers' tool kits and are relevant today, depending on the situation.

Yet another time-honored way to learn about diet is to collect the feces of an animal. Along with stomach studies, biologists working on a wide variety of land animals have relied on analyzing scat samples. The method is appealing because it is non-invasive and includes healthy animals of different sex and age classes. In the early years of whale research, however, the idea of collecting whale poo was not thought to be practical. On the down side was its potential size and the fact that bowel movements happen mainly underwater. Even if they occur near the surface, feces can rapidly sink and be washed away by the ocean. But as researchers have become better acquainted with the whales they study, more research has included the collection and analysis of whale excrement.

One of the fastest of all the cetaceans, a Dall's porpoise in motion is a blur to the naked eye. This photo was taken at a high shutter speed.

and diving behavior is to attach a satellite tag to the whale that may then remain on the whale's body for days or months. The tag allows researchers to track movements through coordinates that are relayed back to a computer. Satellite tags have enabled researchers to define new migration routes and identify previously undiscovered habitats of whales.

While biopsy darts leave no marks on the whale or dolphin, satellite tags are a different matter. Early tags were much larger and more intrusive than those used today, but the current ones can still produce prominent skin injuries. Sometimes, whales and dolphins visibly respond to the darting or tagging and move away or dive deep. Some researchers speculate that the disturbance

for a large whale to darting is comparable to a pinprick for humans, while satellite tagging can leave a deep, potentially long-lasting wound.

All invasive research tools carry risks. Even if the cetaceans appear to be unharmed, disruption to natural behavior and welfare concerns must be taken into account. The starting point for any invasive work should be an examination of the study's objectives. The potential gain must be substantial and should be weighed against the extent of the disruption and the likelihood of success.

A number of questions must be asked:
• How much experience do the researchers have?
• Have the objectives of the research program been submitted to other researchers and wildlife managers for review?
• Have alternative methods been considered?
• Are individual whales and dolphins well known through photo ID or other means? That is, how can

the researcher ensure that individual whales are not targeted more than once?

• Will mothers and calves be targeted?
• Is there a protocol to stop targeting an individual that has been avoiding the researcher's boat for more than 10 minutes?
• Are there systems in place to monitor and report on the welfare impacts on individual whales and dolphins?

For Critically Endangered or Endangered species, obtaining genetic samples or tagging becomes a tricky issue. The need for information on the health of populations must be weighed against the risk of any disturbance to that population, especially mothers and calves.

Of course, biopsy darting and satellite tagging are not the only intrusive techniques. Boats with persistent photographers, filmmakers or scientists determined to get close-up photographs or photo IDs can be even more disturbing to whales.

The use of drones with remote-controlled cameras is another valuable tool. Potentially invasive if flown too close to the whales, drones can obtain data with far less disturbance than from close boat or helicopter approaches if they are used at heights of 175 to 350 feet (50–100 m) or more. Drones have photographed individual right whales for photo ID and have helped determine the health of killer whale individuals with emaciation clearly visible from above. Researchers have collected and analyzed whale spouts to make

Researchers from the Far East Russia Orca Project use drone photography to document a family of killer whales traveling in Avacha Gulf, Kamchatka, Russia. Note the young calf close to mother's side.

health checks. While drones can make a valuable contribution, care must be taken to establish research protocols to plan, conduct and evaluate drone use as part of a research program.

As scientists and humans sharing the planet and world ocean with many other species, we must strive to make only the lightest footprint in the habitats used by wild animals. With around a quarter of all cetaceans in one of the threatened categories, and nearly half rated Data Deficient, whale research cannot be solely for the sake of intellectual curiosity—science for the sake of science. It must have an underlying conservation rationale, especially when invasive techniques are proposed.

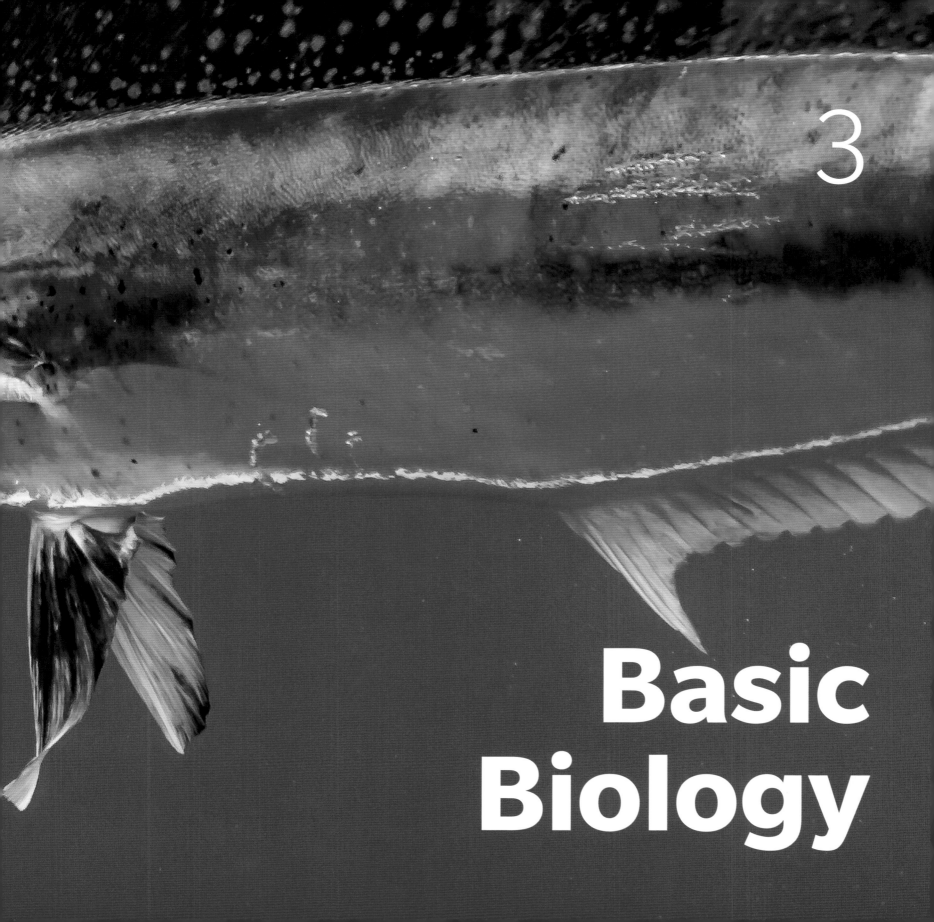

3

Basic Biology

How Cetaceans Function

Whales, dolphins and porpoises—the cetaceans— are social mammals like humans. But in order for cetaceans to be able to feed, breed and travel in water, their external shape and internal biological characteristics have had to undergo substantial changes. Most people know that cetaceans have colonized every ocean from the Arctic to the Antarctic ice caps, but few realize that some species, such as the river dolphins, have evolved to live in estuaries, freshwater rivers and lakes that are hundreds of miles from the sea. Various anatomical adaptations allow whales, dolphins and porpoises to make a living, and many of these adaptations are particular to the depth of ocean or piece of estuary, coastline, river or lake in which they live, and to their choice of diet. Whales strain, skim, gulp, grab, suck or surround their prey. Some whales, porpoises and dolphins hunt in the sandy shallows; others, such as orcas and bottle-nose dolphins, sometimes even slide or roll up on beaches to grab their prey. These adaptations were shaped initially through evolution, followed by cultural exchanges particular to individual cetacean species.

Previous spread: Off Hawaii, a false killer whale prepares to devour a ray-finned mahi-mahi, also known as the common dolphinfish, that is already showing signs of having been attacked. Facing page: A deep-diving Blainville's beaked whale swims up the water column in waters off Hawaii.

Cetaceans belong to the order called Artiodactyla, which some researchers call Cetartiodactyla (comprising the cetaceans and artiodactyls). Inside this order are all of the currently 90 recognized cetacean species—14 are within the group Mysticeti, the baleen whales that strain their food through baleen plates that grow from the roof of their mouth. The other 76 species are the toothed whales, or Odontoceti, which include not only large whales like the sperm whale, but the dolphins and porpoises. The Artiodactyls also include the even-toed ungulates—pigs, ruminants such as cattle, goats, sheep, giraffes, yaks, deer, antelope and the hippopotamus, the most closely related animal to modern cetaceans.

The various species and families of whales, dolphins and porpoises are described and illustrated in later chapters. In summary, the 14 species of baleen whales, the mysticetes, are arranged in four families, and the 76 species of toothsome odontocetes have 10 families. The numerous odontocetes are divided into large-sized toothed whales, which include sperm whales, belugas and narwhals and the beaked whales; medium-sized toothed whales, namely the oceanic dolphins; and small-sized toothed whales, comprising the river dolphins and porpoises.

Baleen whales range in size from the blue whale, the largest animal that ever lived at up to 108 feet 2 inches (33 m) long, though more commonly up to 95 feet (29 m), to the pygmy right whale, which at up to 21 feet 4 inches (6.5 m) is around the average mature size of the largest dolphin, the killer whale.

On the outside, all cetaceans have nostril-like blowholes at the top of the head—two for baleen whales, one for toothed whales. Inside the mouth are varying numbers and sizes of teeth, except for the 14 species of baleen whales whose teeth, after millions of years of evolution, were replaced by keratinous baleen plates that sift the water for food—small shrimplike krill, copepods and other invertebrates, as well as small schooling fishes.

All cetaceans have streamlined bodies, with flippers for steering and tail flukes that they use to propel themselves as they swim. Top speed is an estimated 34 mph (55 kph) for the fastest Dall's porpoises, although common dolphins may be faster. Swimming at up to 23 mph (37 kph), the fin whale is probably the fastest large whale, but even the slowest whales or dolphins swim much faster than humans. Moreover, their stamina is far beyond what humans could imagine. Some gray and humpback whales make journeys of 4,000 to 5,000 miles (6,400-8,000 km) twice a year between breeding and feeding grounds, and some travel much farther. Killer whales have been known to swim from Antarctica to Uruguay and southern Brazil and back in 42 days, a round trip of 5,840 miles (9,400 km). Initial speed was 6.5 knots (7.5 mph/12 kph) but this slowed as the orcas entered warmer waters to 2.7 to 5.4 knots (3–6 mph/5–10 kph). Compare this with human ultra marathon runners. Human marathoners have been known to run 42 marathons in 42 days, an extraordinary human feat. Killer whales might easily run a marathon or two every day in the course of hunting. The swim from Antarctica to Uruguay and back was the equivalent of 5½ marathons per day for 42 days.

Just below the skin of all cetaceans is a thick blubber layer that provides insulation and energy storage. Below the blubber layer, a cetacean's hard and soft anatomy display the skeletal features and organs, heart, stomach and lungs that you would expect to see in mammals. In addition, there are specialized adaptations for cetacean life in the sea. Of special note is its unique respiratory system, which is able to withstand extraordinary pressure changes. Sperm whales have been recorded at depths of more than 4,000 feet (1,200 m), with recorded sustained underwater dives often exceeding an hour.

In 2014, however, after attaching tags to eight Cuvier's beaked whales, Cascadia Research Collective scientists reported new diving records for marine mammals off southern California. These Cuvier's beaked whales achieved record depths of up to 9,816 feet (2,992 m). The whales stayed down for up to two hours, 17 minutes and 30 seconds, also a record for marine mammals. At such a depth, the pressure being exerted is 300 times that on the surface, or 4,380 pounds per square inch (psi). The depth and time spent underwater without breathing are far beyond what a human is capable of, even with scuba tanks. The shallowest porpoise divers put humans to shame. Yet it is the ability to dive deep and return to the surface repeatedly that is the most extraordinary feat.

Whales have special mechanisms to accommodate rapid deep descents and ascents. A whale may start with small amounts of air in the lungs. Below 330 feet (100 m), the underwater pressure collapses the lungs and thorax, preventing gas from entering the blood and reducing the circulation of blood to the muscles. The rapid transmission of nitrogen from the blood to the lungs as the creature returns to the surface may also help prevent problems.

While they typically feed at depths of 1,300 feet (400 m), the deep-diving sperm whales may reach depths greater than 4,000 feet (1,200 m) and stay under for up to 90 minutes.

How Baleen and Toothed Whales Differ	
Baleen whales—Mysticeti	**Toothed whales (includes dolphins and porpoises) —Odontoceti**
14 species; the largest is the blue whale.	76 species; the largest is the sperm whale.
Anatomy	
Baleen plates, two blowholes; overall size larger generally than toothed whales (except for the sperm whale).	Teeth, one blowhole; overall size smaller generally than baleen whales.
Diet	
Mainly krill, copepods and other small invertebrates, small schooling fishes.	Mainly fish and squid, sometimes marine mammals.
Sound	
Low-pitched sounds, broad band to infrasonic; songs in some species.	Broad band to ultrasonic calls and echolocation clicks.
Behavior	
Many migrate from cold-water feeding grounds to warm-water breeding grounds.	Many feed and breed in same general areas, though not sperm whales and perhaps other species too.
Associations, except for mother and nursing calf, are casual. They are much less social than toothed whales, and associations are not long-term.	Many live in long-term groups with the same individuals, but the number and composition varies by species.

Anatomy

On the outside, the body plan of every species of whale, dolphin and porpoise is similar: They all have streamlined torsos with flippers, horizontal tail flukes and blowholes to bring air into their lungs. But with a closer look, the variation becomes more obvious: Some have a dorsal fin, some do not. The flippers are different shapes and sizes—some flippers look more like big ping pong paddles (the male killer whale); others are triangular (the Amazon River dolphin), pointed at the tip (the striped and common dolphins) or rounded (the bottlenose dolphin and harbor porpoise). At up to about 16 feet (5 m) long, the humpback whale's giant winglike flippers are the longest appendages in the animal kingdom. Whale flukes vary in shape too; most species, except for the beaked whales, have a notch in the middle of the trailing edge. Some are almost perfectly triangular with a straight trailing edge, while the humpback has a scalloped, ragged edge.

Skin textures and color patterns vary as well. The big, bulky bowhead whale is a streamlined tank with smooth skin. The closely related three species of right whales have similar size and

Marks of Diversity

Cetacean species differ not just in size but in body shape and markings. The blue whale, top left, has a comparatively small dorsal fin and a long, broad back, with pigmentation that reflects the color of the sky. The Baird's beaked whales, top right, have large beaks and scratched backs. The Risso's dolphin, middle right, has the prominent dorsal fin of most dolphins and a torso that is scratched from teeth marks, while the Risso's dolphin, middle left, shows off scratches and the species' typical blunt nose. Bottlenose dolphins, left, have fewer scratches and a prominent dorsal fin, as well as a beak.

Teeth or baleen—the tools whales use to make a living. Clockwise from top left: the roof of a humpback whale's mouth shows the baleen and fine bristles with which it gathers food; a killer whale has 10 to 13 pairs of teeth on both sides of its upper and lower jaw; each sperm whale tooth is six inches long (15.25 cm); a sperm whale has teeth only in the lower jaw.

Baleen whales have varying numbers of throat grooves, which allow each species to expand its mouth to accommodate food and water. The gulp-feeding humpback, above left, like the fin, blue and minke whales, has many throat grooves. The southern right whale, above right, like other right whales and the bowhead whale, is a skim feeder and has no throat grooves.

bulk but have various roughened patches of skin around the head called callosities. Viewed close-up, callosities look like rough-edged concrete slabs. A number of cetaceans have "piebald" pigmentation, dramatic black with white patches. These include the killer whale, Commerson's dolphin, southern right whale dolphin and Dall's porpoise, but in body size, appearance and behavior, they vary considerably.

Indeed, perhaps the most dramatic difference among cetaceans is in their size. It would take 15 to 20 harbor porpoises laid end-to-end to equal the length of a single blue whale. It would take more than 2,500 porpoises, which weigh up to 150 pounds (70 kg) each, to equal a blue whale's bulk.

Even the blowholes and spouts of cetaceans vary. With two blowholes, the baleen whales produce bushier spouts, while the toothed whales produce thinner spouts from a single blowhole. The blue whale spout is the tallest at 33 to 39 feet (10–12 m). The sperm whale has a blowhole at the tip of its head on the left-hand side; it creates an asymmetrical "leaning" spout that is distinctive even when viewed from miles away. Some of the smaller dolphins and porpoises, as well as the beaked whales, have spouts that are nearly invisible unless viewed close-up in the proper lighting. Spouts—made up of water vapor and nasal mucus, or snot—are easier to see when the air is cold.

If the cetacean rolls over, further clues to the species may be seen. Baleen whales have varying numbers of throat grooves, also known as ventral pleats. The rorquals have many grooves in numbers that differ according to species. The gray whale has only two to five grooves, which are not so much grooves as simple creases. The bowhead and right whales, on the other hand, have no grooves. Many toothed whales may have patterns of white or gray on their underside or no such pigmentation; the pilot whale has anchor-shaped white patches on the underside.

All of these external clues are potentially useful for identifying species at sea. Some of the best diagnostics, however, are hard to spot in the field. More than anything else, the head with the mouthparts, including the teeth, reveal how an animal makes its living, and the way in which an animal hunts, catches and consumes food often provides the clearest identification of species. The baleen whales have no teeth, but the number, size and color of the baleen plates are unique to each of the 14 baleen whale species. The minke whale has the shortest baleen at up to eight inches (20 cm) long, while the bowhead has baleen up to 13 feet (4 m) long.

For a large number of toothed whale species, the number, size and position of the teeth varies considerably and is diagnostic, meaning unique to a particular species. In terms of number, the range goes from no teeth erupted to the 200 to 240 teeth that appear in the long-beaked common dolphin and franciscana.

Teeth are generally conical in the dolphins and spade-shaped in the porpoises. The sperm whale has teeth only in the narrow lower jaw, and these fit neatly into sockets found in the upper jaw.

The most unusual tooth formation appears in the male narwhal as a single tooth, or tusk, that erupts through the upper left jaw and spirals to a length of up to 9 feet (2.7 m). A few narwhals have two erupting teeth, one on either side, though usually the second tusk is shorter.

In some whale species, the appearance and number of the teeth varies between males and females. Most beaked whale males have either two or four teeth that erupt on either side of the jaw, from the middle to the tip of the jaw, often protruding outside the mouth and decorated with barnacles. A special case is the male strap-toothed beaked whale, whose two teeth grow from the lower jaw out of the mouth and around the upper rostrum, preventing the mouth from opening fully. Most beaked whale females, however, have no teeth that erupt. It is thought that beaked whales must suck in their food—squid and fish—and swallow it whole. Whales have three main stomach chambers, similar to their close ungulate relatives, the land-based artiodactyls. But in beaked whales, the pyloric stomach is subdivided further into up to 12 chambers to aid digestion.

With narwhals and the strap-toothed and other beaked whales, the teeth may also be a sexual characteristic, roughly equivalent to the antlers of some of the close relatives of cetaceans on land: deer, sheep and goats. All of these anatomical variations are useful for the identification of species, and sometimes the sex, age and—as we begin to look closer—special pigmentation, marks or scars unique to that individual. These are the keys to unlocking the possibility for detailed studies based on individual photo identification.

Evolution

Whales, like humans, evolved from ocean creatures, whose ancestors lived about 300 million years ago. Today, whales routinely cross the ocean, with some species traveling more than 10,000 miles (16,000 km) a year. But the ancestors of whales also once walked on land. As the potted evolutionary history goes, the whales, unlike ancestral humans, took a look around and didn't like what they saw. About 50 million years ago, they decided to return to the sea. The process of evolving into swimming creatures took about 8 million years, and the details are only now being revealed.

An artist's rendering of *Basilosaurus* depicts a whale ancestor thought to have lived in warm seas all over the world.

Whale paleontologist Hans (J. G. M.) Thewissen, from Northeast Ohio Medical University in the United States, always looks carefully at the ear bones of his finds. No other group of animals has the arrangement of ear bones found in whales, dolphins and porpoises, and it represents an exquisite adaptation to hearing underwater. Even with the early ancestors of whales, according to Thewissen, it is the ear that makes them easy to recognize as such.

Looking at the fossil record, we can see that cetaceans evolved from four-legged, even-toed

An artist's rendering of *Indohyus*, a raccoon-sized distant relative of the whale that walked on land but was adapted to aquatic life.

ungulate ancestors that lived on land some 50 million years ago. From genetic studies, we know cetaceans have DNA similarities to modern ungulates, which include, among others, the giraffe, cow, pig, deer, antelope and hippopotamus. Of those, the hippo is the whale's closest kin. But until the mid-1990s, we had no intermediate fossils that showed how that presumed ungulate ancestor made the huge leap from land to water to become, eventually, the ancestor of the killer whale, bottlenose dolphin, bowhead and blue whale.

In the 1830s, a few decades before the publication of Darwin's theory of evolution by natural selection, large fossil vertebrae were recovered along the riverbanks and in the fields in Louisiana and Alabama. They were initially described as a giant lizard, *Basilosaurus*, but British zoologist Richard Owen recognized that the teeth came from mammals and that the vertebrae were distinctly whale-like. The hunt was on. Collectors soon assembled a 65-foot (20 m) vertebral column with parts of the head and a forelimb.

Basilosaurus turned out to be a very successful 30- to 40-million-year-old whale ancestor. It had short hind legs that could not have supported its weight on land, but it also possessed a tail that was a triangular fluke and hands that were flippers. The fossil finds indicated that these animals lived in all the warm oceans of the world. Whereas the nose opening in modern cetaceans is on the top of the head and is called the blowhole, this structure in basilosaurids was situated halfway between the tip of the snout and the forehead. The eyes were on the side of the head, allowing the animal to see underwater prey. Most important, *Basilosaurus* had ear bones just like modern whales.

The teeth of *Basilosaurus* had a complex shape, unlike those of modern cetaceans and more like those of land mammals. Whereas most modern cetaceans have simple tooth crowns and teeth that look alike from front to back, land mammals, such as humans and also *Basilosaurus*, have four different kinds of teeth—incisors, canines, premolars and molars, and all look different from each other. Yet *Basilosaurus* was too much like a whale to reveal many of the missing steps. Darwin was troubled by the absence of a much more complete whale fossil record. He grappled for harder evidence, and so has every paleontologist since then.

In the early 1980s in Pakistan and later in India, a series of fossil discoveries finally began to fill in gaps. University of Michigan paleontologist Philip Gingerich published his description of *Pakicetus*, based on parts of a skull that included one of the bones of the ear, the tympanic bulla. More *Pakicetus* finds would follow. In 1994, Thewissen, working in Pakistan, announced the first-known pre-whale "with enormous feet" that could walk on land. He called it *Ambulocetus*. Since then, he has led many expeditions to uncover early whale secrets, sharing most recently in the discovery of the pre-archaeocete (or Eocene whale) called *Indohyus* as the oldest extinct species of what would later become the whales, dolphins and porpoises we know today. *Indohyus* was small and land-based, resembling a somewhat chunky mouse deer, one of the smallest ungulates. It weighed less than 18 pounds (8 kg) and its body measured up to 2.5 feet (75 cm). In addition, it had a long, narrow tail. It lived in South Asia during the Eocene period from 46 to 52 million years ago. While it didn't yet look much like a whale, it was on its way. And genetically, it was closer to the whale than to the hippo.

In his 2014 book, *The Walking Whales: From Land to Water in Eight Million Years*, Thewissen traces the path of these whale ancestors—*Ambulocetus*, *Pakicetus*, as well as the even earlier *Indohyus* and others—that had branched off from the artiodactyls, the even-toed ungulates that were once the world's dominant herbivores. Thewissen thinks the raccoon-sized artiodactyls that munched on plants growing along the rivers would hide in the water from danger. Over time, their descendants spent more and more time there, acquiring a physique adapted to swimming fast to catch fish and other prey. Their forelegs thus transformed into flippers, and the simple tails turned into broad, powerful tail flukes used for propulsion. In time, the nostrils moved into a position on top of the head to become blowholes.

The early Eocene whale ancestors, such as *Basilosaurus*, diverged into two lines around 40 million years ago: baleen whales (mysticetes) and toothed whales (odontocetes, which includes the dolphins and porpoises). Early baleen whales still had teeth, and even modern baleen whale embryos develop teeth, although the teeth disappear before birth, at which point baleen starts to form. The toothed whales include most of the modern *Basilosaurus* descendants—currently 76 of the 90 cetacean species.

In spite of the whale's seemingly perfect adaptation to life in water, the legacy of millions of years on land still remain: The whale breathes air and nurses its young, and its paddle-shaped flipper features hand bones with five embedded "fingers." As embryos, cetaceans have tiny hind limbs that disappear before birth, although occasionally, a whale is born with these intact. In the early 20th century, in British Columbia, whalers caught a mature female humpback whale with two 4-foot (1.2 m) hind legs emerging from her body.

How Big Is Big?

Whales are the largest animals that ever lived. First place among living animals goes to the blue whale. At a South Georgia whaling station in 1909, the largest ever female Antarctic blue reportedly topped out at 108 feet 2 inches (33 m), but a more common length for the blue whale is up to 95 feet (29 m). Whales also claim second place among living animals with the fin whale. Southern Hemisphere adults can reach lengths of 88 feet 7 inches (27 m), while Northern Hemisphere adults are less than 79 feet (24 m) long. These are followed by the bowhead at up to 65 feet 7 inches (20 m) and the North Pacific right whale at up to 62 feet 4 inches (19 m). The largest toothed whale is the male sperm whale, which measures up to 63 feet (19.2 m). At least five whale species are larger than the 62-foot (18.9 m) well-named whale shark.

What about the dinosaurs? Blue whales beat most of them in length as well. The recently discovered Titanosaurus from Argentina, however, might have measured 121 feet long (37 m), with an estimated weight of 140,000 pounds (70 tons or 63,500 kg). Thus it would be longer than a blue whale but not quite as heavy.

Maximum blue whale weight is estimated at 397,000 pounds (180,000 kg).

In the sea, certain jellyfish and siphonophore species have tentacles that rival whale lengths, but actual mantle body length is much smaller. Reliable measurements are hard to come by, but the best length record of the giant squid is 39 feet (12 m). Of course, these record specimens would weigh only a tiny fraction of the weight of a single blue whale.

A large whale displays just a fraction of its full length and girth when surfacing, which is why researchers and whale watchers seeing their first whale are often surprised that it does not appear larger. It is only when a whale strands on a beach or a whale skeleton is seen hanging in a museum that its truly staggering size can be appreciated.

But if it's possible for whales to appear small at the surface, the small New Zealand Hector's dolphin and the harbor porpoise can seem

A large whale, like the blue whale pictured at right, displays just a fraction of its full length and girth when surfacing. The much smaller surfacing dolphins, such as the New Zealand Hector's dolphin, above, can seem tiny when they burst out of the water. Both are Endangered species.

tiny. Size is relative, of course, but we humans typically measure size against ourselves. At up to 5 feet 4 inches (1.63 m) with a weight of up to 126 pounds (57 kg), the longest recorded adult New Zealand dolphin is no bigger than an adult human. Its calves are born at a length of about 2 feet to 2 feet 4 inches (0.6–0.7 m). The smallest adult porpoise is even smaller: The vaquita measures up to only 4 feet 11 inches (1.5 m) long.

Migration

Why do the baleen whales migrate? Part of the reason is that conditions in the ocean are constantly changing. When summer comes alternately to the Northern Hemisphere and the Southern Hemisphere, the sun heats the ocean, fueling plankton blooms that produce areas rich in krill, copepods and schooling fishes. These need to be available in large quantities if large whales are going to build up the energy stores to migrate and fast during the breeding season. In winter, the waters are much less productive. But why not eat less, breed and be content to stay near the poles as bowhead whales manage to do in the Arctic? For most baleen whales, migration to warmer waters may provide more protection for breeding, including both calving and raising calves. And they can avoid killer whales and other predators that are generally found in larger numbers in colder waters.

A combination of photo-ID matches and satellite tags have led to the confirmation of new records for whale migrations. It has long been known that gray and humpback whales undertake journeys of 4,000 to 5,000 miles (6,400–8,000 km) twice a year between breeding and feeding grounds, but the surprise has been that some are traveling much farther, although these are not always conventional migrations.

In 2010, a lone female humpback whale swam from her breeding areas in Brazil to the Antarctic feeding grounds and then traveled to a different breeding ground off Madagascar, a long-distance record for a mammal of more than 6,000 miles (9,800 km). Since then, some humpbacks have been recorded swimming from Antarctica, crossing the Equator to spend the winter in the waters off Colombia, Panama and Costa Rica. This would be a round-trip of up to 7,000 miles (11,300 km).

In November 2011, as reported in *Biology Letters* (2015), a satellite-tagged female western gray whale named Vavara started swimming across the North Pacific from Russian waters to the gray whale breeding grounds in Mexico. By the time she returned, 172 days later, she had traveled a total of 13,988 miles (22,511 km). Conventional wisdom had the grays navigating from Mexico to Alaska by hugging the coast all the way, and no one knew where the western gray whales were wintering. Vavara crossed the open ocean and took a different route when she returned to Russia from Mexico. The research also showed that gray whales did not stop much to feed on their long migratory journey, although, on the way back, Vavara did linger along the ice edge in the Bering Sea.

Two more gray whale stories, although not conventional migrations, represent records for long-distance travel. The first was the surprising appearance of a gray whale in the Mediterranean Sea off Israel in May 2010. The photo against the Tel Aviv skyline shocked whale scientists everywhere. Was it a long-lost North Atlantic gray whale? No—that species probably became extinct in the North Atlantic by the 1700s. Instead, this cetacean was an estimated 43-foot (13 m) North Pacific gray that had evidently traveled from the lagoons of Mexico to the feeding areas off Alaska but then had, most likely, headed either east, crossing the Alaskan and Canadian Arctic, or west,

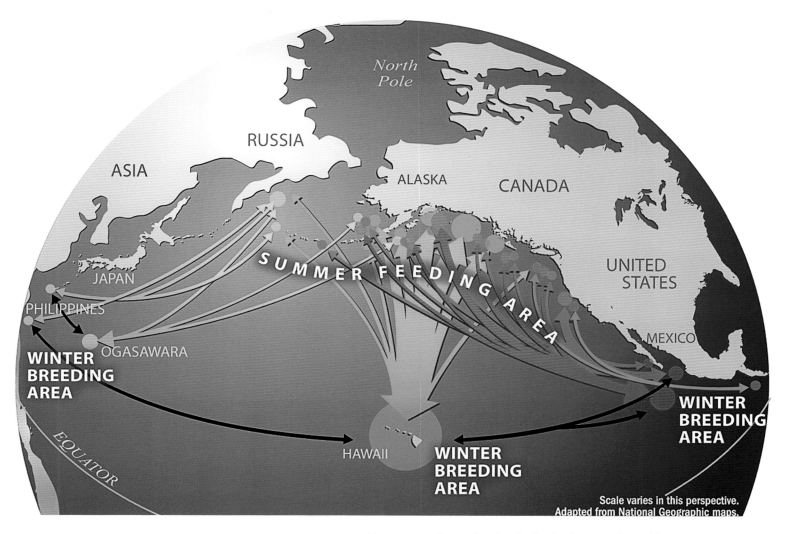

Every year, humpback whales in the North Pacific move back and forth between cold-water feeding areas (in summer) and warm-water breeding and calving areas (in winter). The whales travel on various routes but rarely make straight-line migrations.

across the Russian Arctic. In any case, the easterly direction through the Northwest Passage had become, by 2007–2008, ice-free. Once through that passage, the whale might have rounded eastern Canada, or perhaps Greenland, and then headed south until it sensed the warm, salty waters of the Mediterranean. In search of the usual winter habitat in the lagoons of Mexico, the whale may have turned east through the Strait of Gibraltar, traveling to the other end of the Mediterranean before whale researchers noticed.

This whale could not be matched to the Mexican gray whale catalog, but researchers concluded

that, measured point-to-point, a one-way journey of at least 12,993-miles (20,910 km) had been undertaken. The same gray whale was seen three weeks later off Barcelona in the western Mediterranean. Since then, this adventurous gray has disappeared.

In May 2013, a different gray whale turned up even farther away from Mexico. Dolphin tour

Learning from Stranded Whales

Long hours on the water with calm seas, lots of spouting whales and the chance to learn something new—that's the starting point for a whale biologist's perfect day. When you study any of the some 22 species of beaked whales, however, chances are you'll be dealing with a dead whale on a remote beach rather than with a live specimen. It's the only opportunity scientists have had to study some beaked whale species.

Whales, dolphins and porpoises strand along rocky shorelines and beaches in many parts of the world. As the tide recedes, the cetacean is left high and dry on a beach or lodged on rocks. This is when researchers, local authorities or members of the public see the whale, dolphin or porpoise—if they don't smell something first. When the cetacean is still alive, there may be the urge to try for a rescue. What is the best course of action if you find a stranding?

The first caution about any cetacean stranding is that members of the public should not touch or try to move a stranded animal. Seek expert help immediately. Most places in the world have a protocol for how to handle strandings of marine mammals. Expert help can be found through local, county, council, state or provincial government offices, a natural history museum or local police station. There are often out-of-office 24-hour numbers.

Strandings must be divided into two main categories: individual or mass, followed by four further subdivisions of the above: live or dead, and large or small. If the stranded cetaceans are alive, there is the question of whether there should be an attempt to return them to the water or whether they should be euthanized. The course of action should normally only be decided and undertaken by the national or local authorities, with a veterinarian skilled in dealing with stranded cetaceans. Individual strandings often involve sick, dying or dead animals. If they are alive, there may not be much hope of successful rescue or return to the sea. Sometimes, the death may be natural mortality. It could also be a result of entanglement in nets, a ship strike or a disease that may have come from pollutants. In any case, the issue of the animal's welfare, alleviating suffering, should be paramount.

The course of action also depends on the size of the individual cetacean that is stranded. It will not be easy

Facing page: In one of the largest mass strandings in New Zealand's history, some 400 long-finned pilot whales beached themselves at Farewell Spit in February 2017. Volunteers raced to the scene to take part in an attempt to save at least 100 of the whales. In an all-too-common scenario, most of the rescued whales re-stranded themselves hours after being refloated.

to move a mature killer whale that weighs several tons back into the water, and the same is true of any size humpback, fin, blue, sperm or any other of the large whales. In these cases, if the whale is still alive, the most humane action may be to put the animal down, if that is possible, or to help make them comfortable following the directions of the researcher, veterinarian or authority who is in charge. Still, some large cetaceans have been successfully returned to the sea.

Strandings are sometimes the result of a natural situation, such as a storm at sea, unusually high tides or water movements, a whale's search for food or an attempt to evade predators. Harbor porpoises have ended up on beaches after attacks from bottlenose dolphins. Mammal-hunting killer whales have attacked seals and sea lions on dry land, sometimes stranding themselves in the process, although they are usually skilled at returning to the water. Gray whales feeding in the shallows close to shore along the U.S. and Canadian Pacific coasts are rarely in any danger of stranding.

Mass strandings are a different situation, and in many cases, the cause cannot be determined. Perhaps disease or parasites have led to malnutrition. Sometimes only one individual in the group is ill or

disoriented, but the others refuse to leave. In other cases, the cause leads back to the harmful pollutants or algal blooms that have moved up the food chain via ingested shellfish. More sinister are situations that result from trauma or disorientation following navy sonar or loud seismic exploration noises. Several beaked whale species, especially the Cuvier's beaked whale, have been documented as victims in the Canary Islands, Greece and the Bahamas.

How to respond to a stranding

• Keep people and dogs away; don't touch the cetacean(s).
• Make a quick assessment, take notes and a few photographs. Note the following: Alive or dead? Any injuries or blood? Are there distinctive markings on the back or dorsal fin? Estimate the size, and identify the species, if possible. Note time of day, location in relation to landmarks and tide (GPS, if possible). Note how accessible the area is for vehicles.
• Report the stranding using your notes and photographs by calling authorities (local, county, council, state or provincial government offices, natural history museum, or police station), and follow their instructions.

boats in Walvis Bay, Namibia, first reported the sighting, which was confirmed a week later by the Walvis Bay Strandings Network. It was the first known record of a gray whale in the Southern Hemisphere. Not only was this gray whale in the wrong hemisphere, but it was also in the South Atlantic.

No firm conclusions can be drawn, but these sightings have stimulated considerable scientific discussion and prompted a series of questions:

• Are we witnessing some of the first results from global warming and the melting of the ice?

• To what extent are whale distributions already changing?

• If whale populations continue to recover, is this a return to how whale populations were once distributed throughout the oceans? Is it also a preview of distribution in the millennia to come?

Baleen whales are the big migrators, the long-distance truck drivers of the ocean, but not all baleen whales undertake migrations across hemispheres. The Bryde's whale spends most of the year in subtropical waters, undertaking shorter offshore migrations. One population of humpback whales, the Endangered Arabian Sea humpback, also lives year-round in warm subtropical waters. Certain blue whales, too, may feed and breed in the same tropical area around the productive Costa Rica Dome. Meanwhile, the bowhead whale, with its thick blubber layer, prefers cold waters year-round, responding to the advance and retreat of the ice. These short-haul species move within their respective zones, responding to changes in prey distribution and life history, courting, mating, calving and raising calves.

Other long-distance migrators break the rules too. Fin whales in the western Mediterranean spend their year modestly moving between the Ligurian Sea off Italy and France to warmer seas off North Africa in winter. Some gray whales migrating along the Pacific coast of North America don't make the entire Mexico to Arctic waters migration but stop off and spend the summer along the coasts of Oregon, Washington and British Columbia, where they find the small shrimplike crustaceans known as mysids and other food along the muddy shore.

As ocean temperatures change and polar regions open up due to global warming, marine mammal scientists wonder if migrations will change. In recent years, researchers have made an effort to learn more about migratory behavior, and there have been many surprises.

Toothed whales do not migrate like baleen whales. Many toothed whales live and feed year-round in the same general area, while others travel when needed in search of food-rich areas, ranging over wider areas. That may include inshore summer and offshore winter movements or other patterns related to the temperature and productivity of the ocean. Sperm whales are a special case: Some bulls live in cold temperate and subarctic waters while females and young stay in the tropics and subtropics. Occasionally, the males make journeys to the warmer waters to mate. Killer whales are constantly on the move in search of prey, typically traveling distances of 75 to 100 miles (120–160 km) per day. But the record killer whale swim from Antarctica to Uruguay and southern Brazil, and back, a round trip of 5,840 miles (9,400 km) in 42 days is equivalent in distance to many baleen whale migrations. No one knows when, how often or even precisely why killer whales might travel so far.

Social Behavior

Whales, dolphins and porpoises love to socialize:

• A blue whale gulping krill seems to be feeding all alone along the north shore of the St. Lawrence as the first snow of winter starts to fall. An hour later, researchers from a high vantage lookout see several more blues feeding in the same fashion for several miles all along the coast. The blues are all within communication distance of each other.

• Off Maui, Hawaii, a big humpback female slaps her 16-foot (5 m) flipper against the back of a male, making her presence felt. Half a mile away, and easily audible underwater, a group of young males sing their hearts out.

The foundation of cetacean associations is the mother-calf bond, as can be seen in the physically intimate connection between these humpback whales on the tropical breeding and calving grounds.

• In Patagonia, Argentina, an older killer whale teaches two youngsters to beach themselves in preparation for the season when sea lion pups appear on the beaches.

Whales, dolphins and porpoises, as with most other social mammals, are usually found in groups, commonly referred to as pods or schools and sometimes herds. There are also specific names used for groups within certain species: matrilines, communities, clans, associations and others. Some of the large whales spend time on their own but eventually join up for group activities such as feeding, migrating or breeding. Solitary-seeming individuals may also be in acoustic contact. As noted above, blue whales need space to feed, and other blues may not be obviously associated, but an aerial flight or hydrophones often confirm that other blues are rarely more than a few miles away.

The big baleen whales travel in pairs, threes, occasionally fours and sometimes up to 10 individuals. Killer whale pods can contain fewer than six individuals (the mammal-eaters) or up to 15 to 20 (the fish-eaters); community aggregations can be 100 or more. By comparison, some of the oceanic dolphin species, such as common dolphins, may travel in groups of several thousand individuals. Sometimes these traveling or feeding groups even contain individuals of several dolphin species.

The foundation of cetacean associations is the mother-calf bond. Whales, dolphins and porpoises are born fully formed with vision, hearing and the ability to swim. Still, they are vulnerable, culturally ignorant of the ways of their population and species. There is a window of only a few minutes after birth before a calf needs to breathe. With various dolphin species, another member of the group pushes the baby to the surface to take its first breath. The fetal folds, showing as unevenness in the skin, may still be visible. The dorsal fin may lie flat. But within a day, that baby is taking milk from its mother, the dorsal fin stands up, and the calf is swimming like a pro.

Most whales, dolphins and porpoises stay with their mothers for one to two years, and the mothers focus their energy on protection and teaching life skills to their calves. The calves of fish-eating North Pacific killer whales, however, spend their entire lives near their mothers. When female calves come of age and begin having their own offspring, they may start to form their own sub-unit, but they still remain close to their mother. But mature orca males spend the span of their full lives, about 30 years, traveling beside their mother, with only short breaks when they go off and mate after encountering killer whales from their community with different dialects.

Family associations among whales vary in terms of the kind or degree of closeness and the longevity of the relationships. The bond may be loose and temporary or it may be permanent. It may be cemented along matrilineal lines so strong that only death can sever the connection.

The big baleen whale mothers have a year or two with the baby in tow on migration, but then they sever the ties. Even so, on the feeding grounds, some whales associate with one another. The spectacular groups of humpbacks blowing bubble clouds and surfacing together suggest that there is some benefit in looking for food together and feeding side-by-side. The advantages are more evident in the systematic group hunting of killer whales, which together create waves that can knock seals off ice floes. In the first example, the humpback whales may or may not be related, but the orcas have matrilines or extended families that know and live with each other full time.

With the well-studied fish-eating killer whales, the entire matrilineal group includes males,

A tight group of false killer whales allows a bottlenose dolphin to join up with them in the waters off Dominica in the Caribbean. Most dolphins are so social that in the absence of conspecifics, they will join up with other species either temporarily in order to hunt or feed or more permanently for social reasons.

Previous spread: In the waters off Dominica in the Caribbean, multiple mothers in a typical mixed group of female and young sperm whales take turns babysitting the young while others dive deep for food. Above, a newborn humpback whale calf is pushed to the surface for a first breath and glimpse of the world above.

Why do cetaceans stay close together? The early-life mammalian connection to the mother starts to explain it, but why do they continue to remain together? The standard explanation is that higher numbers of animals help ward off killer whales and sharks as well as offer support in hunting by corralling food and sharing knowledge about food sources. But every species has its own strategies. Small groups might assist in catching limited predictable prey, while larger groups may be organized to catch the more abundant prey that appears from time to time, as well as to protect the young and fend off predators. Being part of a large group also reduces an individual's chance of being eaten by a large predator.

females and young, all of which have a relationship with a central older female. With sperm whales, the matrilineal society includes just the mothers and their calves and siblings. Older males form separate groups that only visit the female-led groups from time to time for mating.

In contrast to these stable groups with fixed membership, there are fluid fission-fusion societies in which the members change over time. Even so, some of these societies have at least a few members that form stable alliances, particularly the males. This is true of the bottlenose dolphin, the northern bottlenose whale and the Baird's beaked whale, among others. The long-term associations may be for as little as one or two years, but there is limited data from which to draw more concrete conclusions.

The striking evidence of the social connections, particularly in the toothed whales, becomes evident when you take a close-up look at individuals. In some species, scars and scratches appear literally all over the body. A few of these scars are from predator attacks, parasites or nets, but most are from members of their own group. Older individuals, especially the mature males, appear to have the most scars. These could be partly the result of play, and they hint at complex relationships. A standard explanation is that male alliances may help in the herding of females and keeping them away from other males, but there is not much evidence of this. Instead, researchers speculate that these intricate social relationships represent a built-in community network for spreading knowledge within a group.

Courtship and Reproduction

In late autumn in Roseway Basin, off the southern tip of Nova Scotia, Canada, a big female North Atlantic right whale—nearly 50 feet (15 m) long and weighing some 70 tons (63,500 kg)—splashed beside our ship, but her presence had nothing to do with us. Some whales interact with boats, but our vessel did not interest this female. Not this time. She had love on her mind—or she was playing "hard to get." Seconds later, she issued a loud bellowing call easily heard through the boat.

Right whales vocalize, but it appears that they do not sing songs. They are one of the few baleen whales, along with the gray whale, whose vocalizations do not fit with our human definition of songs, which for various animals, from birds to whales, have been defined as sounds, or sequences of sounds, that are repeated. The role of songs in the courtship and mating process is still being worked out with the humpback, bowhead, blue and fin whale males that sing songs on the breeding grounds.

But with the right whale, it appears that no one bothers singing. We can only speculate: With her loud call, was the female beside our boat actually signaling her readiness to the males? In other whale species, as researcher Bill Watkins has noted, the males make the displays toward females. Why would the right whale do things differently?

Across the cold, flat, calm expanse of ocean, we could see areas of disturbed water in the distance, like a line of surf or an advancing swell. Then, from various points on the horizon, we glimpsed the approaching male right whales. More questions: If the female was calling the males to come to her, then why, as they approached, did she invert her massive torso, pushing her belly out of the water, thus making mating difficult?

The males arrived and stationed themselves around the female. With his enlarged, outstretched penis, one tried to enter her at an impossible angle. Eventually, she had to roll over to spout and breathe, and that's when the action really got underway. One by one, the males positioned themselves to try to enter her. There was jostling for position but no fighting. After as many as five or six matings by different males, the deed was done.

What do scientists make of this? Researcher Moira Brown says that that this breeding strategy means that the right whale with the most or strongest sperm succeeds in getting his genes into the next generation. It's sperm competition. Thus, through evolution by natural selection, the right whale penis and testes have become the largest in the animal kingdom, with the penis on average 7 feet 6 inches (2.3 m) long and both testes topping out at a ton (900 kg).

Right whales—including North Atlantic, North Pacific and southern right whales—are more surface-active in their mating, which is why we know more about them than we do about most other whales. Right whales engage in sexual play throughout the year, but successful reproduction usually happens in winter.

The nuts and bolts of courtship and reproduction vary, as far as we know, among the 90 species of whales, dolphins and porpoises. Still, there are some similarities. They all mate belly to belly. The females nearly always have only one calf,

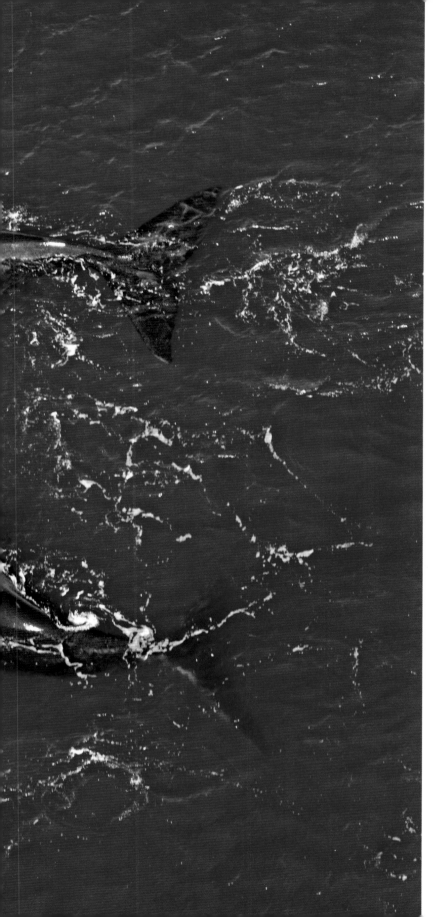

though, rarely, twin fetuses have been reported in some baleen whales, sperm whales, belugas and a number of dolphin species, including bottlenose and common dolphins. There are even cases of Siamese twin stillbirths with humpback and sei whales. No whale twins are known to have survived.

The big migrating baleen whales, including the humpback, right and gray whales, focus their breeding activities in the winter months. Mating periods vary with the dolphins and porpoises—some of them are recorded as year-round, although many have seasonal peaks. An exchange of vocalization is often part of the courtship, though the role of song in humpback and other large whales is still being debated. Gestation periods vary from 10 to 13 months for most baleen whales, while toothed whales vary from 10 to 17 months. Squid-eating whales, such as the sperm, pilot and beaked whales, as well as the Risso's dolphin, have a longer gestation period (11 to 17 months), compared with the fish-eaters (10 to 12 months). British researcher Peter Evans suggests that the diet may influence the rate at which the fetus can grow in the embryo, and that the longer period may be because cephalopods, such as squid, have a lower energy content than most fish and plankton.

About a year after mating in Roseway Basin, the big female North Atlantic right whale, swollen and heavy with calf, lumbered into the warm, shallow waters off the southeast United States, between the Carolinas and northern Florida. To reach the

Off Península Valdés, Argentina, multiple male southern right whales scramble for mating position with a female. The most persistent males are the most likely to get their genes into the next generation.

A southern right whale male prepares to mate on the breeding grounds near Hermanus, South Africa. Pursued females, after announcing their readiness to mate, typically roll over and, belly up, play hard to get.

young and have a period of a few months to several years raising the calf and resting between calves, with the males having little or no role. Some of the toothed whales are known to nurse for up to three to four years, although 18 to 24 months is more common. Age at sexual maturity varies from harbor porpoises (3 years); minke whales (males at 3 to 6 years; females 5 to 7 years) and gray whales (5 to 11 years for males; 8 to 12 years for females). Fin, sei and Bryde's whales reach sexual maturity at about 10 years. Bowhead whales reach sexual maturity at age 25 years. Most North Atlantic right whale females have their first calves when they are between 10 and 14 years old, and the average calving interval is three to five years.

The minke whale and harbor porpoise can breed every year, even with their 10-month gestation period, but for most cetaceans, the breeding cycle is two years or more. The cycle for the blue, Bryde's, humpback, sei and gray whales is 11 months of gestation, followed by a lactation period of six to seven months and a six- to seven-month resting period for a minimum two-year cycle. Bowhead and right whales have a three- to four-year breeding cycle. The baleen whale cycle is synchronized to fit in with migration.

The well-studied female fish-eating killer whales are unusual in many respects. These killer whales have calves only every five years. The female calves stay beside their mothers for up to 15 years, while the males stay their entire lives, up to 50 years.

sanctuary of those protected waters, she had to dodge ship traffic and fishing nets. The 13-foot (4 m) female baby, once expelled, was pushed to the surface for her first breath. Cetacean birth size varies from 2 feet 4 inches (70 cm) for a newborn harbor porpoise to 23 to 26 feet (7–8 m) long for a newborn blue whale.

In the right whale's warm protected waters, the calf can nurse on mother's rich milk, which is 30 to 60 percent fat and 5 to 15 percent protein, as opposed to human and cow milk, which is 2 to 4 percent fat and 1 to 3 percent protein. Nursing lasts at least six months before mother and calf part company in the northern feeding grounds later that year.

Most female whales and dolphins suckle their

Whale, Dolphin and Porpoise Life Histories

Cetaceans take their time to mature. Then the females put their energy into raising calves with every individual being considered important. Yet among the various species, there are considerable differences in terms of age at maturity, calving interval and longevity.

Species name	Female age at maturity	Gestation period	Calving interval	Female longevity	Notes
North Atlantic right whale	10–14 yrs	12 mos	3–5 yrs	70+ yrs	
Bowhead whale	25 yrs	13–14 mos	7 yrs	>115–130 (200+) yrs	May be longest-living mammal
Gray whale	6–12 yrs	11 mos	2 yrs	75–80 yrs	Average age of maturity is 8 yrs
Humpback whale	5–10 yrs	11.5 mos	2 yrs	>50 yrs	Calving interval is sometimes 1 yr
Blue whale	5–15 yrs	10–12 mos	2–3 yrs	80–90 yrs	
Fin whale	7–8 yrs	11 mos	2 yrs	90 yrs	
Sperm whale	9 yrs	14–16 mos	5 yrs	>50 yrs	
Northern bottlenose whale	7 yrs	12+ mos	2 yrs	40–50 yrs	
Baird's beaked whale	? yrs	c17 mos	?	54 yrs	Males live longer, up to about 84 yrs
Beluga	9–12 yrs	14–14.5 mos	3 yrs	40+ yrs	Some evidence longevity could be 80+ yrs
Bottlenose dolphin	5–13 yrs	12 mos	2–3 yrs	>57 yrs	Sexual maturity in males is 9–14 yrs, and they live up to 48 yrs; females up to 48 yrs have given birth
Killer whale	10–15 yrs	15–18 mos	5 yrs	>80–90 yrs	The North Pacific fish-eating orcas have five calves over 25 yrs, then 10+ yr. menopause; males live on average about 30 and up to 50 yrs
Harbor porpoise	3–4 yrs	10–11 mos	1–2 yrs	10–24 yrs	

Whale Longevity

What is cold water's secret power? In 2006, Bangor University scientists studied 400-year-old Arctic clams (*Arctica islandica*) in the icy waters off northern Iceland and found a clam that was 507 years old. It was later dubbed the Ming Clam, because it had started its life during the Ming Dynasty.

In 1999, biologist Craig George's team from the North Slope Borough Department of Wildlife Management in Barrow, Alaska, attempted to age bowhead whales using amino acids in the lenses of the eyes of animals taken by the Native Alaskan hunters. He produced several dozen age estimates for adults, mostly between 20 and 60 years old, but four of the whales were estimated to be over 100 years old; one was thought to have reached 211 years of age.

These results were controversial because they exceeded all the known ages of whales and other mammals. In 2007, however, George examined a bomb lance fragment, the point of a harpoon, which had been recovered from the right scapula of a 49-foot-long (14.9 m) male bowhead that had just died. Working with John Bockstoce from the New Bedford Whaling Museum, George determined

Removed from the neck of a bowhead whale in 2007, this harpoon fragment had a long relationship with its target: It was patented in 1879.

that the big male must have carried the harpoon point in his body since he was unsuccessfully hunted in the late 1800s. Calculations based on when this type of harpoon was in use (it was patented in 1879) indicated the whale would have been 115 to 130 years old at the time he was killed.

In an effort to gain insights into the evolution of longevity, researchers led by João Pedro de Magalhães of the University of Liverpool, U.K., sequenced the bowhead whale's genome. It was the first large whale genome to be sequenced. The team announced their results in *Cell Reports* in 2015, and the work was trumpeted as potentially applicable to the fight against age-related diseases in humans. Although they have a thousand times more cells than humans, bowheads do not appear

to be more at risk of cancer. Perhaps there are natural mechanisms present in bowhead whale cells to suppress cancers? The researchers also identified differences in the bowhead genome when compared with smaller mammals, including a much lower metabolic rate and changes in one gene involved in thermoregulation that may be related to the metabolic differences in whale cells. Research is in the early stages.

But it does raise the question: What is it about Arctic waters and longevity? In addition to the bowhead, at least one narwhal has been aged at 115 years.

The research on the bowheads, narwhals and clams has all involved dead individuals or their tissue. On the other hand, outside of the Arctic, a female killer whale in J pod named "Granny," a southern Vancouver Island orca community member that died in late 2016, was thought to be more than 100 years old. She was the matriarch of this community of Endangered killer whales, and her age was estimated through her progeny and social group, although genetic research indicated her age was probably mid-60s to 80s.

Songs, Calls and Clicks: Unraveling the Sonic World

The deep, resonant songs of whales and the higher-pitched chatter of dolphins and porpoises were a feature of the ocean long before humans walked on Earth.

Thanks to the 1970's best-selling album *Songs of the Humpback Whale*, which followed the discovery of humpback songs by Roger Payne and Scott McVay, the humpback whale's songs are the best known. Indeed, the humpback is the symphonic composer of the whale world—in terms of what humans can hear, its songs are by far the most prevalent songs in the ocean. Up to 30 minutes long, these songs feature various themes or movements sung in predictable sequences, from low, rumbling parts to trumpet-like passages that reverberate across the ocean floor. The sounds span seven octaves, nearly the range of an entire 88-key piano keyboard. During the winter mating season, they are repeated over and over for hours at a time. Around Hawaii in the winter months, researchers who drop a hydrophone into the ocean almost anywhere are guaranteed of tuning into a whale song—or several. The songs are so loud that they can sometimes be picked up through the boat hull even without a hydrophone.

Researcher Katy Payne and other researchers have dissected the songs from different ocean basins. The typical humpback song has eight themes that are sung in order, with the last theme running back into the first. Within each theme are common phrases. Each humpback has some improvisational latitude in terms of the number of phrases that can be sung. Sometimes, they may sing a phrase only twice; other times, they might repeat it 15 or 20 times.

Only male humpback whales sing, and all the males on the same mating grounds essentially sing the same song, though the songs change gradually through the winter breeding season. At the end of the season, the song is put into storage as the whales migrate and focus on feeding all summer. However, in autumn, while the humpbacks are still on the feeding grounds and it is months before the peak of migration, some singing occurs—perhaps this is the whale version of tuning up. After migrating back to the mating grounds, the whales appear to pick up the song where they left off; after two to 10 years, themes are either substantially changed or even eliminated. The new themes are based on new phrase types that come into the song.

Worldwide, the rules for humpback songs are largely the same, but the content and speed of change are different. In the Southern Hemisphere, the themes change much more often. Researcher Michael Noad was recording humpback song off Australia's east coast in the South Pacific when he detected a strange humpback song. Within the space of a season, the other whales began to imitate it. This particular song was produced by an Indian Ocean humpback that may have altered his migration route. Noad, Ellen Garland and others later traced the spread of humpback song across the South Pacific over three years as it moved steadily from the eastern Australia breeding grounds to French Polynesia.

According to researcher Jim Darling, song themes persist for years in the North Pacific. During the winter, there is evidence that songs on the breeding grounds off Hawaii and off Mexico

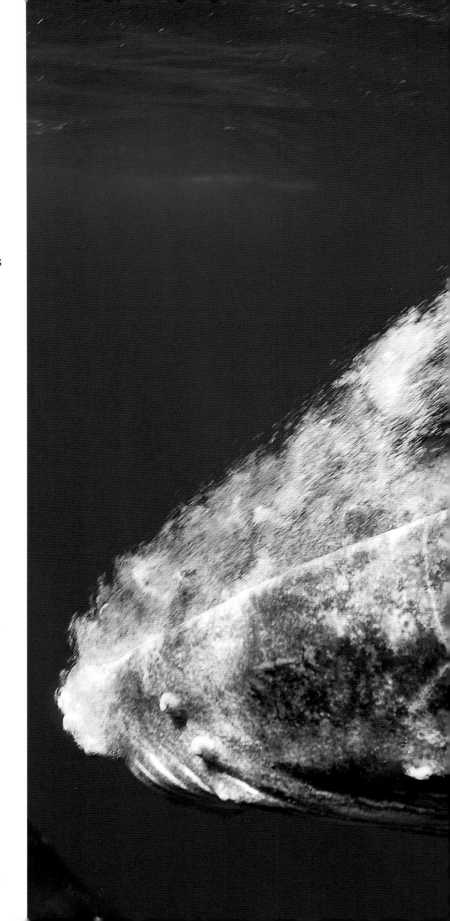

diverge slightly. Yet over several years, the songs converge and thus remain similar, even though these areas are 3,000 miles (4,800 km) apart. In a comparison of songs from Hawaii to winter breeding areas in the Philippines and Japan, the overall patterns and phrases were the same, but the Hawaii humpbacks used some additional unique phrases. Researchers Hal Whitehead and Luke Rendell suggest that the North Pacific whales may have more exchanges on migration than in the South Pacific due to the geography of the North Pacific landmasses—the northern feeding areas are relatively close together between Russia and Alaska. Humpbacks do sometimes sing on migration. Some whales even travel between breeding grounds, which increases the potential for song sharing.

Over several decades, Darling has worked around Hawaii trying to determine why male humpback whales sing such complex songs. The hypotheses are still flying around. While the general agreement is that the songs have to do with mating, there is no evidence that the songs attract the females. Are the songs dominance displays—the equivalent of underwater antlers? Perhaps the song is a male-to-male signal that indicates the status of the males. But the similarity of the song produced by all the males and the fact that it leads to brief, friendly interactions among them seems to argue against these ideas.

Darling's current hypothesis is that the song provides an index of association, a means of organizing long-term relationships among males, and that this may be a key to understanding the

A humpback calf in the waters off Tonga blows a stream of bubbles— good practice for various feeding maneuvers in later life.

Highly vocal common bottlenose dolphins have been the subject of study in cetacean communication and echolocation since the 1940s.

mating system. Off eastern Australia, researcher Joshua Smith and others speculate that the songs stimulate or prepare the females, the goal being to get them in the mood for possible mating. Whitehead and Rendell conclude that either or both of these hypotheses, as well as others, could be true or not. Darling admits that, more and more, he suspects that "we are missing something major." In any case, for Whitehead and Rendell, the complex, beautiful and ever-evolving humpback songs provide some of the best evidence that whales have culture.

Other whales sing songs, according to the definition of repeated sequences of sounds heard on the mating or breeding grounds, but these songs sound monotonous compared with humpback whale songs. Known singers are the minke, fin, bowhead and blue whales. The Bryde's,

sei and pygmy right whales may sing, while gray whales and the various species of right whales are definitely not singers.

Bowhead whales have units within a few phrases that are repeated through the winter breeding season. Like humpback songs, bowhead songs cover a wide frequency range and they change through the season. There may be two or three songs being sung on a breeding ground, each of which changes completely from year to year. Bowhead songs are far less complex with the cycles lasting no more than a minute. Some researchers compare them to human folk songs whereas the humpbacks compose symphonies; others caution that this interpretation could easily be wrong.

To humans, the simplest-sounding songs belong to fin whales. With a simple one-second dip in frequency, they are at the low edge of human hearing. To hear them properly, it is necessary to speed up the recording. Only males make the sounds, which are loud and travel for many miles across the ocean. Their sounds differ in some areas, and in the future, this fact may allow us to identify distinct populations or breeding units.

Blue whales grab their share of attention with the lowest of all known songs, lower even than fin whales and louder—they can be heard thousands of miles away. As with fin whales, blue whale songs have only one repetitive theme. Blue whale themes have one to five different units. Below human hearing, as well as below the range of most audio speakers, blue whale songs can only be heard by humans if we speed up the recording.

Researchers have found that blues can sing for days, taking breaks only to surface and breathe. They have found 11 different song types around the world that may correspond to distinct populations of blues. From year to year, the song structure generally stays the same, with one important difference. Researcher Mark McDonald and his colleagues have found that, year by year, the songs of blue whales off Australia, as well as in other areas of the world, have decreased in pitch, or frequency, becoming deeper and deeper. The whales that live off California, for example, now sing at a frequency 31 percent lower than they did in the 1960s. It may be a mating strategy—how low can you go—if females like deep voices.

Other hypotheses for blue whales decreasing their frequency include the need to avoid interference with the slightly higher fin whale songs or the growing numbers and sizes of blue whales in those populations that are starting to recover.

And what about the levels of increasing noise in the ocean? What if a blue whale wants to be heard by a potential mate, or by competing males, over the longest possible distance? Noise has been known to cause other species to change the volume or frequency of vocalizations, but if that were the case, blue whale songs would be going higher in order to avoid chronic ship noise.

The most intriguing speculation comes from Whitehead and Rendell in *The Cultural Lives of Whales and Dolphins*. They say that blue whale song culture may have changed in a particular direction for no particular reason, but once established, the conformist nature of the songs leads the declining-pitch trend to spread around the world. They compare it to hemlines of women's skirts going up and down over the decades. If "cultural drive" is the reason, then eventually, the frequencies just might go back up. Since whale songs are recorded from year to year, researchers will eventually be able to test some of these theories.

While the baleen whales are the singers of the cetacean world, many of the toothed whales, including the dolphins and porpoises, produce echolocation clicks, part of the sonar system they use to find their food and their way through dark seas. Typical echolocation consists of the dolphin or other toothed whale sending out a stream of clicks that bounce off an object and return as echoes which are read and computed as sound pictures. As they hunt for giant and other squids, sperm whales are the master echolocation practitioners. Dolphin sonar has been shown to be a precision tool that is able to detect size, shape, texture and movement, including minute differences in the objects that would be undetectable to the human eye. But sperm whale echolocation clicks are lower than those of dolphins and 20 decibels louder—on the order of 100 times louder.

The Value of Identifying Killer Whale Dialects

Killer whale acoustic researcher Olga Filatova, from Moscow State University and the Far East Russia Orca Project, is a good listener. She is one of only a handful of scientists who would score high marks if there were a global orca dialect exam. In such an exotic test, the examiner would play a short recording of orca sounds from any number of places around the world: Russia, Alaska, British Columbia, Iceland, Norway, Argentina, New Zealand and various sites around the Antarctic. The blindfolded researcher would have to determine where the sounds came from—one point would be awarded for identifying which part of the ocean, one point for the precise orca clan, and a bonus point if the precise pod can be named.

Orca dialect competitions might one day happen—but voice and other sound recognition software, used to identify human recording artists and songs, for example, will probably get there first. For now, the value of being able to identify dialects is substantial. It means that killer whale pods can be identified at certain locations and times of year wherever hydrophones record the sounds. These recordings can be live, allowing researchers to follow up with sightings and further investigation, or they can be made through the use of a fixed recorder that is retrieved periodically. The whales can be recorded at night, during storms and through the winter, at times when researchers are fast asleep or stuck in their labs.

Filatova started recording Russian killer whales off Kamchatka in 2000, revealing repertoires of stereotyped calls—vocal dialects specific to each pod. This finding aligns with John K. B. Ford's discovery of dialects in the 1980s in the northern and southern community Vancouver Island killer whales off the west coast of North America. The Russian sounds from the western North Pacific were different from those that Ford found. In Southeast Kamchatka alone, Filatova has identified three acoustic clans: the Avacha clan (at least 13 pods), the K19 clan (at least two pods) and the K20 clan (at least five pods). The Avacha clan is the most common in Avacha Gulf, but in the Commander Islands, the K19 and K20 clans are encountered more frequently.

Filatova has investigated how sounds depend on behavior and social context. Based on the observed differences, she found that biphonic and high-frequency monophonic calls are mainly used as unit and pod markers and to help track the position of pod members when they are widely separated. Low-frequency monophonic calls, on the other hand, are used as intra-group signals to maintain close-range contact between pod members.

In other studies, Filatova and her colleagues have looked at the cultural evolution of vocal dialects in killer whales. Orca dialects are an example of cultural change accumulating slowly over many generations. The dialects are learned at birth from the mother and other members of her matriline. Over decades, it was thought that some of the call types in the dialects change due to random learning mistakes and innovations that accumulate through time, so that more related, or recently diverged, families have more similar dialects. However, Filatova and her colleagues found that dialect similarity is not always connected to relatedness, i.e., learned mainly from the mother. This may be due to constraints in the structure of the calls that lead to their convergence in unrelated families; it also may be due to the borrowing of calls from other families.

Studies show that killer whale sounds depend on behavior and social context.

4

Baleen Whales

Right and Bowhead Whales · Family Balaenidae

North Atlantic Right Whale

Eubalaena glacialis

Reaching and occasionally exceeding lengths of 59 feet (18 m) and weighing up to 99 tons (90,000 kg), right whales are so named because whalers considered them the "right" whales to hunt. Originally considered one global species, they are now divided into three: the North Atlantic, the North Pacific and the southern right whales. Centuries of whaling, first in European waters and then in the coastal waters of North America, drove the North Atlantic right whale close to extinction, and there has been no whaling of this species since 1935. With only about 500 individuals, the North Atlantic right whale remains on the U.S. Endangered Species List. The species was rediscovered in the Bay of Fundy in the early 1970s. Since then, the species has received the attention of the scientific community in an effort to determine why it hasn't recovered.

Right whale researchers Scott Kraus, Amy Knowlton, Moira Brown and their colleagues found an answer by proving that right whales were getting hit and killed by ships as fast as they could reproduce. In the cold, food-rich waters in and around Stellwagen Bank National Marine Sanctuary, at the mouth of Massachusetts Bay, right whales spout and splash, chugging back and forth, rippling the water from just beneath the surface with their bulky bodies and wide, paddle-shaped flippers. Swimming with their mouths open, these sturdy whales skim-feed through dense patches of copepods, gorging themselves on their primary food.

All around the whales, ships come and go from Boston Harbor. Some approach dangerously close to the whales. What is clear is that this marine sanctuary is not a sanctuary from ships. With input from scientists working with the sanctuary, a so-called traffic separation scheme (TSS) has moved the ships into a lower-density whale area. Still, the ships sometimes hit whales.

Much can be explained by the fact that right whales are oblivious to the world when they travel, feed and engage in courtship activities. They are often close to the surface of the sea. And unlike killer or humpback whales, which have a more flamboyant appearance at the surface, right whales have a low profile, which adds to the likelihood of their being struck by ships.

Based on this research, Defenders of Wildlife, The Humane Society of the United States, Ocean Conservancy and Whale and Dolphin Conservation went to court in 2008 to demand that speed limits be put in place. Speed limits in certain whale areas were cut to 10 knots (18.5 km/h), and in 2012, these groups petitioned the U.S. government to extend those protections indefinitely. In 2013, National Oceanic and Atmospheric Administration researchers found that imposing these speed limits on ships passing through or close to the 10 North Atlantic right whale seasonal management areas along the U.S. east coast had reduced the risk of ship strike by 80 to 90 percent.

North Atlantic right whales are also threatened by fishing gear entanglements that can trap, seriously injure and often kill them. According to the National Marine Fisheries Service, the agency charged with protecting right whales in U.S. waters, the rate of serious injury and mortality from fishing gear entanglements is more than three times what this species can sustain.

Marine protected areas and other area-based protections can work effectively if there are good management plans, enforcement and monitoring. In 2009, the Center for Biological Diversity, joined by the original groups that had gone to court in 2008, petitioned the U.S. government to expand federally designated critical habitat for this species. After years of delay, in 2016, they reached a legal settlement—an agreement resulting in additional protected habitat designated for the species. Now that scientists have identified North Atlantic right whale critical habitat, there's a chance to save an Endangered species.

Previous spread: A group of feeding humpback whales charge to the surface, mouths wide open. Above, a humpback whale off Hawaii has become fatally entangled in fishing gear.

Conservation Status

Scientists have evaluated many of the world's marine and land-based species as part of their work for the Species Survival Commission within the International Union for Conservation of Nature (IUCN). The IUCN is the umbrella body of nature and conservation organizations and world governments and regularly asks for conservation advice from the expert scientists in the Species Survival Commission.

For whales, dolphins and porpoises (the cetaceans), the relevant IUCN Species Survival Commission group is the Cetacean Specialist Group, about 100 scientists from around the world. This group has given a rating for each cetacean species and in some cases for subspecies or populations within a species. The ratings are as follows: Not Evaluated, Data Deficient, Least Concern, Near Threatened, Vulnerable, Endangered, Critically Endangered, Extinct in the Wild, Extinct. Only Least Concern is considered a favorable rating.

The ratings for each species appear on the IUCN Red List (iucnredlist.org). In this book, whenever a species rating is mentioned, it is shown with the first letters capitalized, for example, "Critically Endangered."

North Pacific Right Whale

Eubalaena japonica

At up to 62 feet 4 inches (19 m) in length and weighing in at up to 99 tons (90,000 kg), the North Pacific right whale is the largest of the right whale species. It looks much like the North Atlantic right whale, though a little stockier. With broad, paddlelike flippers and deeply notched tail flukes, it is mostly black in color with occasional white patches. As with the other right whales, there are callosities on its large head, and it lacks a dorsal fin or dorsal ridge on its smooth, broad back. Like the North Atlantic right whale, the closely related North Pacific right whale is an Endangered species—thought to number around 500 individuals. Yet much less is known about this whale.

Following heavy whaling in the 19th century, North Pacific right whales were fully protected after 1935. Notwithstanding that protection, Russian whalers secretly killed at least 681 more whales in the 1950s and 1960s.

The discovery of two dozen right whales in feeding areas in the western Gulf of Alaska and the southeastern Bering Sea led the United States to declare these areas as critical habitats in 2006. There have been only rare sightings in all other areas in U.S., Canadian and Mexican waters. Recently, on the Russian side of the Pacific, there have been more sightings, and the fate of the whales in this region may determine the future of this species.

The Far East Russia Orca Project, Russian Cetacean Habitat Project and other research teams working in the Russian Far East and northern Japanese waters have started a North Pacific right whale catalog with photo IDs of all of them.

A southern right whale plies the waters off Patagonia.

From 2003 to 2014, there were 19 encounters of 31 whales, most of them after 2009. Two of these were of two whales, and one sighting was of three and another five. Every sighting of a right whale provides a little more hope, but encounters with multiple individuals, especially females with calves, shows that social bonds may still support a future for this species.

Southern Right Whale
Eubalaena australis

Like the North Atlantic and North Pacific right whales and the bowhead, the southern right is an impressive size, reaching lengths of some 55 feet 9 inches (17 m) and weights of at least 88 tons (80,000 kg). And like these other big whales, the southern right whale has no dorsal fin or ridge. The head is similarly marked with callosities, and the mostly black body typically features patches of white on the stomach and chin.

In 1971, Roger and Katy Payne set up the first long-term wild-whale research program at Península Valdés, Argentina, focused on southern right whales. Studies continue today, with researchers from Ocean Alliance and Instituto de Conservación de Ballenas, led by Mariano Sironi, Victoria Rowntree and others. Using aerial photographs of the individually distinctive patterns of callosities on each whale's massive head, they have photo-identified more than 3,000 individuals. Following the lives of these known individuals has allowed these researchers to determine important biological variables. For example, females have their first calf at nine years of age; afterward, they have one calf every three years, on average. This information has been used in models to

A southern right whale surges out of the water, showing off its diagnostic head callosities, white stomach patches and black flippers shaped like near-perfect trapezoids.

understand the population's size and growth rate.

In addition, by knowing "who's who" in the population, researchers have also been able to chart behavioral differences as well. Mochita and Rombita are female right whales that were born in the same gulf of Península Valdés in 1999. From a young age, Mochita was observed to be an extroverted individual that spent time socializing with other whales, while the introverted Rombita preferred to be alone and avoided interaction.

Using photo ID, researchers are also learning about habitat preferences. Hueso was born in 1999. The researchers saw her again with her first calf in Golfo Nuevo, on one side of Península Valdés, in 2006, and then with another calf in Golfo San José, on the other side, in 2014. Hueso used both gulfs as nursery areas, while other whales might use just one. Some whales, like Troff, the first whale ever to be identified in 1970, use completely different calving grounds in different years: Troff gave birth to three calves in Golfo San José in the 1970s but then had another calf in 1988 in southern Brazil, 1,860 miles (3,000 km) farther north. She returned to Brazil in 1994 but was back in Golfo San José, at Península Valdés, in 2004.

The individual differences in behavior and habitat preferences reveal much about the whales' lives as well as their conservation needs. Thanks to Troff and other whales photo-identified on Antarctic feeding grounds, researchers now know that conservation strategies to protect right whales in the South Atlantic can only be effective if they include large oceanic areas that extend from southern Brazil to Antarctic waters. Southern right whale numbers are growing at a rate of 5 to 7 percent per year. If that continues, the population could double every 10 to 12 years. That's good news, especially when compared with the forecast for the closely related North Atlantic right whale species, whose habitat is near large cities and ship traffic along the U.S. and Canadian east coasts. The number of whales caught in nets or annually hit by ships has been erasing any population increase. By contrast, southern right whales are doing well. Still, there are problems.

Around Península Valdés, Argentina, whale deaths have increased since 2005 from an average of 6.5 per year to about 65 per year, 90 percent of which are calves less than three months old. The main cause may be biotoxins from harmful algal blooms that are ingested and have been known in the past to kill even very large whales. Other factors that are contributing to the deaths could include food supplies that are affected by ocean warming; infectious disease; and kelp gulls that harass calves by gouging out pieces of their backs and feeding on their skin and blubber. Southern right whales still have a long way to go before they return from near extinction.

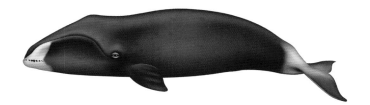

Bowhead Whale
Balaena mysticetus

The fate of the "ice whale"—most closely associated with Inuit hunters across the Alaskan (U.S.) and Canadian North—is changing. Heavily whaled in the Atlantic region over three centuries from the 1600s, the slow-swimming, massive bowheads are still killed in the western Arctic by indigenous subsistence hunters, though at a comparatively low level. Overall, their numbers have been on the rise. Research shows that bowheads, which reach lengths of up to 65 feet 7 inches (20 m), may live from 115 to 130 years, if they are lucky enough to have dodged more than a century of harpoons. Some may even live to 200 years of age.

The bowhead occupies a unique niche in the ocean. It is the only filter-feeding baleen whale resident in Arctic waters. Unlike most other baleen whales, it does not migrate to warm temperate or tropical waters but spends its life moving with the ice.

A skim-feeder, the bowhead swims along, mouth partly open, and food and water enter through a gap in its baleen—at up to 13 feet (4 m) long, the longest baleen of any whale. The baleen fringes catch the food as the water passes through the baleen slats, which grow from the margins of the mouth. The bowhead's closest whale relatives, the three right whale species, also employ this skim-feeding strategy. Like the right whales, the bowhead eats copepods and krill (euphausiids).

Researcher Bernd Würsig and his colleagues often watched bowhead whales from a low-flying airplane off Alaska. They witnessed the whales skim-feeding in a V-shaped echelon formation, with up to 14 whales staggered behind and to the side of one another. The bowheads were moving in the same direction and at the same speed, and Würsig's team hypothesized that such coordinated movement increases the efficiency of feeding on concentrations of small invertebrates.

As befits a cold-water mammal, the bowhead has a thick blubber layer, indeed the thickest of any whale species, and a somewhat triangular head that occupies a full one-third or more of its box-shaped body. As it travels along the icy water, its shiny black torso barely skims the surface, and if it is not too windy, its tall, bushy, hot-breath spouts can be seen from some distance, backlit by the low Arctic sun. Sometimes it dives under the ice, where it is able to locate breathing holes and fissures. While it does not possess the precision echolocation used by dolphins and sperm whales, it may employ a kind of passive sonar by listening to the reflected echoes from its low moans, which sound different under ice than they do in open water. The tapering head and overall bulk may allow it to break through ice up to almost three feet (60–90 cm) thick.

Individual identification is more challenging with whales that have a low profile and no dorsal fin. Even so, a bowhead that spyhops or rolls on its back near the water's surface reveals dark spots that are arranged like a necklace on the white chin. From the air, variable white chin patches on the protruding lower jaw, as well as other marks and scars on the body can be seen. Through dedicated study, researchers have been able to build up bowhead whale photo-ID catalogs as well as conduct aerial surveys to estimate numbers and assess distribution. There are five main populations worldwide, the three largest of which were last estimated at a total of about 24,000, still lower than the original population that thrived before whaling. These three larger populations are rated Least Concern on the IUCN Red List, but the other two are considered Endangered. The United States has kept the bowhead whale on its Endangered Species List.

Today, the bowhead faces new threats. With global warming, ecologists and conservationists wonder to what extent Arctic conditions may change. Will the opening of the Arctic as the ice floes melt drive temperate species to invade these waters? Will killer whales turn up in much greater numbers and start chasing bowhead calves? Will the growth in shipping, oil exploitation with increasing noise levels, and incidents of ship strike pose more problems? Will the secrets to longevity in cold northern waters disappear with climate change? Or will new habitats open up that may benefit the whales? We don't know.

The bowhead may be the first test case for global warming in large whales. On the plus side, we know that the bowhead must have survived some warm periods in the past, but climate change is happening at a faster rate now, and historic periods did not include widespread human use of bowhead habitat. Its year-round Arctic residence leaves this whale fewer options than more wide-ranging species. It has nowhere else to go.

A rarely seen pygmy right whale shows signs of scarring, possibly caused by cookie-cutter sharks.

Pygmy Right Whale · Family Neobalaenidae

Pygmy Right Whale
Caperea marginata

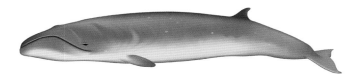

The pygmy right whale is not related to the right whale. The resemblance is limited to its somewhat arched jawline. At up to 21 feet 4 inches (6.5 m) long, this whale is hardly a pygmy either, although it is the smallest baleen whale. Rarely seen, the pygmy right whale is probably often mistaken in profile for the Antarctic minke whale. If the short flippers or blunt-head and arched jawline are visible, identification is more likely.

Resident to the colder waters of the Southern Hemisphere, the pygmy right whale is usually seen alone or in pairs. Researchers in South Australia have reported assemblages, perhaps mass social occasions, with more than 100 pygmy right whales occurring in one group.

Blue Whale

Balaenoptera musculus

Slender and streamlined, the rorquals are distinguished from the right whales by their long, narrow flippers and the dorsal fin, positioned about two-thirds along the back. Another shared characteristic is a series of longitudinal skin folds called throat grooves, which start at the mouth and expand to allow the whales to accommodate huge mouthfuls of water and food before the water escapes through the baleen plates.

Most well-traveled whale researchers and whale watchers have a favorite story about the largest animal in the world—the blue whale. One of mine is the sight of seven blue whales spouting around our ship in front of gleaming Snæfellsjökull, the Icelandic glacier mountain made famous in Jules Verne's *Journey to the Center of the Earth*. We were five hours out from Olavsvik on Iceland's north-west coast, and the seas had calmed when the blue whales appeared. Two of them were young blues—always a good sign for an Endangered species. With an estimated population of only 1,000 to 2,000 blue whales in the central and eastern North Atlantic, the appearance of seven meant that close to 1 percent of the population had gathered around our ship.

I have also encountered blue whales off Quebec, eastern Indonesia and Baja California, Mexico. For pure pleasure, few days at sea can compare with watching blue whales in the Gulf of California. This is one place to celebrate the existence of blue whales—the eastern North Pacific is the only blue whale population in the world known to be steadily increasing. Before the whaling era, more than 300,000 blue whales are estimated to have traversed the world ocean, most of those around the Antarctic. Today, there may be as few as 10,000 widely dispersed, potentially searching for mates. Some estimates range as high as 25,000 for all populations and forms of blues, some of which, like the pygmy blue, may well turn out to be separate species.

The range of the North Pacific blue whale extends from southeast Alaska, along the west coast of Canada, the United States and Mexico to the offshore waters of Costa Rica. Small parts of its habitat are protected in three U.S. national marine sanctuaries along the California coast—the Gulf of the Farallones, Monterey Bay, Cordell Bank and Channel Islands national marine sanctuaries. Despite having a massive broad head, the streamlined blue is fast-moving, but that does not prevent it from being hit by ships as it feeds in the productive waters just offshore from busy southern California ports. The expanding shipping industry and increasing blue whale numbers in the eastern North Pacific mean that unless its

A blue whale in the clear waters of Baja California submerges after spouting. Catching a ride are three remoras near the tip of the rostrum, feeding off ectoparasites and loose skin flakes on their host.

habitat is specifically protected in the shipping lanes, the blue whale will continue to be struck and killed. Blue whale collisions are also common in the shipping lanes around Sri Lanka.

One probable breeding area for the blue whale is the Costa Rica Dome/Papagayo Upwelling, in the Eastern Tropical Pacific. This productive area is unique in that the blue whale not only breeds here but also comes to feed. Some of the California and Mexican blue whales are known to travel here. In 2009, Whale and Dolphin Conservation (WDC) proposed the "Dome" as an ecologically or biologically significant area (EBSA) to the Convention on Biological Diversity (CBD). In 2014, the CBD formally approved this EBSA as special year-round habitat for the blue whale and various tropical dolphins, sharks and sea turtles. A Costa Rica conservation group named MarViva, together with WDC and other groups, are now working on a formal marine protected area (MPA) designation that would include parts of national waters of Costa Rica and Nicaragua as well as the high seas outside the 200-nautical-mile (370 km) zone.

Perhaps more than almost any other whale, the blue whale can claim the world ocean as its home. In the early 1990s, Cornell University biologist Christopher Clark was allowed to eavesdrop on blue and other whales using the U.S. Navy's North Atlantic network of hydrophones for submarine detection. Clark detected blue whales sending their deep booming sounds across the ocean basin. More

A blue whale, facing page, swims through the waters of the Channel Islands National Marine Sanctuary off California. Inset: The massive skeleton of a blue whale is the centerpiece of many national museums. Here, we see the blue whale's size relationship to a human.

A blue whale lifts its tail flukes in preparation for a feeding dive.

recently, some of these sounds have been classified as songs, and researchers have found that there are as many as 11 different types of songs corresponding to various ocean regions.

Scientists disagree over how many discrete groups or populations of blue whales there are in the world, but most recognize a separate Antarctic subspecies, a pygmy blue whale form in the southwest Indian Ocean, as well as the blue whale in the Northern Hemisphere, which is generally smaller than the Antarctic subspecies. In 2006, acoustic researcher Mark McDonald and his colleagues, drawing on more than three decades of data, proposed that each of the 11 different song types should represent a separate breeding unit because songs are normally used in a mating context.

Why did it take so long to discover the song of the blue whale? Although these songs are extremely loud, they are emitted at such a low frequency that, for the most part, audio speakers are incapable of reproducing them. Most are below the level of human hearing. In order for humans to appreciate these songs, they must be recorded on special recorders, then sped up four times during playback. Even sped up four times, blues still have deep bass voices.

Fin Whale

Balaenoptera physalus

As a group of fin whales swam in a clockwise circle in the northern Gulf of St. Lawrence, Canada, their powerful, sky-high spouts broke the stillness, and then, one by one, the surfacing whales rolled to the right slightly, as if to hide. Fin whales have asymmetrical pigmentation, with the lower right-hand side of the head, the edge of the upper lip and the first third of the baleen colored white or pale gray. The left-hand side is dark brownish gray to black, as is the top of the head, back and flanks. No other whale species looks like this, except for the much smaller Omura's whale. Why?

Early in the study of fin whales, researcher Edward Mitchell and his colleagues tried to test the theory that the fin whale employs a special right-side-down lateral lunge feeding strategy to maintain counter-shading, that is, to present white areas so as not to alert prey. Instead, Mitchell found that the other large baleen whales, including blue, sei and Bryde's whales, often fed right side down as well.

Nevertheless, these color patterns are valuable to scientists for identification purposes. On the right side, the pale

A fin whale displays this species' asymmetric right lower chin and lips. The left-hand side, not seen here, is uniformly dark.

Asymmetry in the Whale World

Much of nature appears to have a remarkable symmetry: In the cetacean world, every species has two eyes and two flippers on either side of the head, the tail fluke and the dorsal fin or ridge is centrally positioned. But the asymmetries are also remarkable:

• The fin whale is dark gray or brown to black on its back and flanks and white on the underside. However, on the lower right-hand side of the head and baleen, extending to the upper lips along the rostrum, the coloring is white instead of dark. The left-hand side of the head is the same uniform color as the rest of the back and flanks. But there is a symmetrical twist to this asymmetry: The asymmetrical coloration is reversed on the tongue.

• Most toothed whales have a blowhole positioned slightly to the left of the center on top of the head. The sperm whale's blowhole is positioned not only far to the left but is angled forward. When big male sperm whales spout, the blow is slanted such that it seems that the whale must be tilting his head forward and to the side.

Male narwhals have a tusk (which is actually a tooth) that penetrates the left-hand side of the upper lip and keeps growing. Defying asymmetry, some narwhals grow two tusks on either side—but it is extremely rare, and the right-hand tusk is smaller.

• The narwhal's tusk is a tooth that penetrates the upper left-hand side of the mouth, spiraling out to nearly 9 feet (2.7 m). Some narwhals have two tusks, and for them, things are more symmetrical, except that the left-hand tusk is typically longer than the right-hand tusk, and both tusks have counterclockwise spirals. The tusk is a male characteristic, but in rare cases, females have one tusk or, in extremely rare cases, two.

The sei whale, shown above, has symmetrical head coloring and one ridge on its head. In size, it is larger than the Bryde's whale and smaller than the fin whale.

Fin whales in the open ocean undertake migrations across ocean basins that span great distances. But what happens when fin whales live in a relatively closed sea like the Mediterranean? Researchers led by Simone Panigada and Giuseppe Notarbartolo di Sciara from the Tethys Research Institute in Italy have spent two decades trying to piece together this fin whale puzzle. First, they used photo ID to distinguish individuals, establishing that the whales were resident through the summer in the Ligurian Sea. Researchers sought habitat protection, and in 2000, France, Italy and Monaco agreed, creating the Pelagos Sanctuary for Mediterranean Marine Mammals. But were whales coming and going through the Strait of Gibraltar to the open North Atlantic? Occasional winter sightings south of Sicily were tantalizing. In 2015, satellite tags revealed the path of the whales out of the sanctuary as they moved south toward the North African coast. These fin whales were full-time Mediterranean residents.

pigmentation extends back to form what is known as the blaze, a light V-shaped chevron behind the blowholes that goes down each side, the precise pattern of which, along with dorsal fin shape and scars, makes these whales individually identifiable, even if the degree of variation is subtle. Some whales in every population are nearly impossible to distinguish.

After the blue whale, the fin whale is the second largest whale, and it has songs with the second lowest frequency. Like the blue whale, it is also Endangered, though not as endangered, at least in terms of world numbers. Fin whales may be the fastest-swimming large whale, but they were not quick enough to avoid whalers' harpoons. As with many of the large whales, the body of biological information on this species first came from the whaling ships and the researchers who examined their hard and soft anatomy and reported on their findings. The largest fin whale measured was 88 feet 7 inches (27 m). Now we are learning more about what the fin whale is like in the wild.

In the Mediterranean, fin whales are being struck and killed by ships and entangled in fishing gear. Although it is more difficult to document, they may also be affected by noise, prey depletion and pollution. One of the busiest areas of the ocean, with extensive coastal developments, the Mediterranean has 30 percent of the world's ship traffic on 1 percent of the ocean's surface area. How we address human impacts on whales and the marine environment in this ancient sea in the middle of Earth may well predict our chances of success elsewhere.

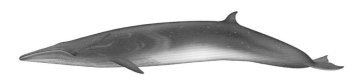

Sei Whale

Balaenoptera borealis

The gray back was splotchy, covered in small circular scars left by lampreys and cookie-cutter sharks. This whale was a big female, though not the record size of 59 feet 1 inch (18 m). She was skimming, not gulping, her food. Surrounded by minke whales just outside Skjalfandi Bay, Iceland, I suddenly realized the whale spouting next to our boat was too big to be a minke whale. She was not big enough to be a blue or a fin whale. Instead, she turned out to be my first sighting of a sei whale, probably a female because they are typically larger than the males. Seconds later, she was gone—the sei whale is known to reach speeds of over 15 mph (25 km/h).

The least known of all the big baleen whales, the sleek sei whale has often been confused in scientific accounts with the Bryde's whale and sometimes with the fin whale. A single ridge on top of the rostrum (compared with three ridges in Bryde's whales) and symmetrical head coloring (compared with the asymmetrical white chin on the fin whales) is diagnostic—unique to this

At right, a Bryde's whale breaches in the new Swatch of No Ground marine protected area in the Bay of Bengal off Bangladesh. Overleaf: A Bryde's whale prepares to dive into a bait ball containing thousands of Pacific mackerel.

In the sequence above, a Bryde's whale devours small schooling fish off South Africa, its throat expanding to take in prey and water.

species—but hard to glimpse unless seen close-up.

To feed, the sei whale usually skims the water for copepods and other smaller fish prey, although it sometimes lunge feeds. This distinguishes the sei from the other rorquals, which are solely lunge-and-gulp feeders; the most notable skim feeder on copepods is the right whale. In comparison with the right whale, the sei is a much faster swimmer.

In the commercial whaling era from the 1950s to the 1970s, the sei whale was exploited heavily as blue and fin whale numbers declined. It remains Endangered today, and researchers have yet to make detailed studies of the whale that shuns coastal waters and spends most of its life on the open ocean.

Bryde's Whale
Balaenoptera edeni

The Bryde's whale is the tropical counterpart of the bowhead whale—preferring to do its breeding and feeding around the same latitudes—although like the bowhead, migrations may occur over shorter distances. These whales tend to congregate around known food-rich areas of the tropics, such as the Gulf of California, the Caribbean and the Gulf of Mexico, although some groups move seasonally. Bryde's whales off South Africa, for example, move to equatorial West Africa in winter.

In the past, the Bryde's whale profile and spout at sea were easily mistaken for that of large minke whales or small sei whales. Closer up, however,

the three ridges running lengthwise on top of the head are diagnostic—unique to this species.

Bryde's whales may be one or several species. The main Bryde's whale has a worldwide tropical and subtropical distribution. In addition, there are smaller forms that tend to be more coastal in distribution. These may be separate populations, subspecies or even species.

The whale species called **Omura's whale (*Balaenoptera omurai*)**, once thought to be a pygmy form of the Bryde's whale, is now recognized as a separate species. This species has a fairly restricted distribution, as far as is known, from southern Japan and southern Korea in the north to the Indonesian archipelago and to Papua New Guinea in the south. The Omura's whale does not appear to reside in the mainland coastal waters of China or Southeast Asia. Though not closely related to Bryde's whales, it is a similarly tropical and subtropical rorqual.

Common Minke Whale

Balaenoptera acutorostrata

The smallest of the rorqual whales, an adult common minke whale is about the size of the largest dolphin, that is, an adult male killer whale.

In the 1980s, the late American whale researcher Eleanor "Ellie" Dorsey cracked the minke whale photo identification code. She had worked with Roger Payne on southern right whale photo IDs and had a knack for identifying whales; an early right whale catalog was referred to as "Ellie's bellies." Later, Dorsey moved to the U.S. Northwest and decided to work with minke whales. She found some minke whale hangouts in Puget Sound, but

the minkes were fast, elusive and hard to get close to, much less photograph. The main photo-ID work going on in the area then was of the killer whale, identified by the dorsal fin shape and fin markings, and the humpback whale, identified by the patterns of black and white on the underside of the tail. The problem was that minke whales rarely flipped their tails, and the dorsal fins seemed to be mostly the same. In conversations with Richard Sears, who was then trying to develop a method to identify individual blue whales, Dorsey determined that the pigmentation on the whale's back was unique from whale to whale. In a sharp photograph taken in good light, the pigmentation was distinctive. Dorsey and colleagues created the first minke whale catalog and over several years showed that these minke whales had home ranges and were returning to the same areas year after year to feed.

Minke whale habits vary considerably from place to place. In many areas around the world, they seem to keep to themselves. Yet so-called friendly minke whales approaching or following boats have been documented in the Gulf of St. Lawrence, Canada, around Iceland and off northeastern Australia, among other places.

In recent years, the waters around Iceland have offered some of the best whale watching of minke whales and other whales in the world. Unfortunately, minke watching has been compromised by whalers who persist in killing minke and fin whales using Iceland's self-allocated whaling quotas. Whale watchers eager to see whales in the wild have watched in dismay as whalers

Researchers off eastern Australia have been able to identify individual dwarf minke whales by photographing them underwater.

harpoon minke whales. They've also witnessed the spectacle of dead fin whales, their huge bodies strapped to the side of a whaling vessel, as they are transported to the fin whaling station for processing. After receiving many complaints, Iceland's government set aside a separate near-shore area for whale watching and an adjoining area farther out to sea for whaling. Still, whalers continued to catch minkes close to, or even inside, the designated whale-watching area. There were fears that the whales were being driven away by the whalers or worse, showing curiosity toward the whale-watching boats and then having this "friendliness" exploited by whalers. As of 2016, after a decade when 706 Endangered fin whales were killed around Iceland, the fin whale hunting has ended for now, with the whaling company blaming the collapsing Japanese market for fin whale meat. But minke whale hunting continues.

By contrast, off northeastern Australia, in the Great Barrier Reef, a unique population of minke whales, a subspecies known as the "dwarf minke whale," has developed a long-term relationship with researchers. These researchers have documented the underwater behavior of these whales and have developed citizen science programs to engage the public.

In the Southern Ocean, especially around Antarctica, a new species of minke whale was declared recently, separate from the common minke whale, called the **Antarctic minke whale (B. bonaerensis)**. This species is being hunted by the Japanese in their "scientific whaling" industry, which operates outside the best advice of scientists from around the world. For minke whales, whaling is not a thing of the past but a constant hazard faced in the waters of three oceans: the

An Antarctic minke whale feeds near the ice shelf of Antarctica—one of the most productive marine areas in the world. The Antarctic minke whale, which can be up to 35 feet 1 inch (10.7 m) long, was recently declared a separate species from the common minke whale.

North Atlantic around Iceland and Norway, the North Pacific around Japan, and the Antarctic and Southern Ocean.

Because of its comparatively small size, the minke whale was the last whale to be commercially hunted, with some 284,000 killed. Today, between one and two thousand are killed every year by whalers from Japan, Iceland and Norway —admittedly a small number when compared with the wholesale slaughter in previous centuries of blue, fin, sei, right, gray, humpback, bowhead and sperm whales. Yet these minke whale operations contravene the spirit of the 1986 worldwide whaling moratorium. The Antarctic minke whale is rated Data Deficient on the IUCN Red List List, while the common minke whale is rated Least Concern.

Humpback Whale

Megaptera novaeangliae

A humpback whale is far from being an icon of whale beauty. While it may represent an evolutionary branch of the sleek, streamlined rorqual group of whales, its anatomy is taken to some bizarre extremes. The flippers appear to be too long and ungainly for the up to 55-foot-9-inch-long (17 m) body of the whale to support—even though researchers have found that they are hydrodynamic and highly adapted for maneuverability. The humpback knob of a dorsal fin, sitting on a humped platform on the back, looks like a gnarled appendage next to the elegant crescent fins of other whales and dolphins. It has to be said that the humpback is aptly named.

And then there's the humpback face. Baleen whales are known to have tiny hairs, mainly along their jawline, but the humpback is the only whale to have tubercles, or bumps, located all over the head. These tubercles are in fact huge hair follicles. Every humpback has between 30 to 60 tubercles, each with a sensitive hair that may be used to detect the movement of prey or to enhance the sensitivity of humpback-to-humpback contact. Perhaps the hairs are left over from the humpback's land-based ancestors. No one knows.

When it comes to whale song, however, the humpback more than makes up for its looks. The monotonous strains found in the baleen whale repertoires of the blue, fin and minke are primitive plainsong compared with the symphonic complexity of the operatic humpbacks.

Field researchers at sea, as well as whale watchers, are finding it increasingly easy to find the humpback whale these days:

• Along the British Columbia coast, longtime whale-watching skipper Jim Borrowman notes that only a few decades ago, there were no humpback whales to be found. The same is noted by Vladimir Burkanov, Alexander Burdin, Olga Filatova, Olga Titova and Ivan Fedutin around Kamchatka and the Commander Islands of Russia. Now summer-feeding humpbacks return in greater numbers every year.

• Off Southeast Alaska, researcher Fred Sharpe has observed two fundamentally different kinds of humpback whale feeding behavior. The first is conventional feeding on various prey, including krill. Humpbacks may feed close together, but typically, they hunt on their own for whatever is available. Some humpbacks, however, are determined to catch and eat the wily herring and to do so have formed a kind of herring-appreciation collective, assembling in tight groups and working together to corral the fish. Some specialize in a feeding strategy that features bubble nets, which are created when the whales blow underwater bubbles that rise to the surface, thereby confusing, herding and entrapping fish schools. Others make loud sounds, while still others whack the surface with their flippers. The whales then take turns swimming up through the bubble nets and lunging for the herring. All of this noisy, heated

action can attract other humpbacks. The amazing thing is that the newcomers seem to catch on quickly, helping with the herding and taking turns devouring the nutrient-rich herring.

• In New England, researcher Mason Weinrich watched as "lobtail feeding" in one individual humpback spread to more than 40 whales over a decade. Lobtail feeding is enhanced bubble-cloud feeding. With this feeding strategy, a whale lifts its tail and then slams it down on the surface one to three times in an effort to herd sand lance (also known as sand eels). Then the whale returns to bubble-cloud feeding. The tail-slamming component may have evolved after a humpback's chance discovery that it led to a bigger catch; perhaps other humpbacks, whether or not they were related to each other, saw it and imitated it.

Gray Whale · Family Eschrichtiidae

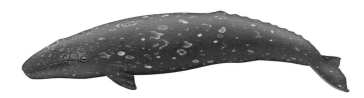

Gray Whale

Eschrichtius robustus

At up to 49 feet 3 inches (15 m), the gray whale lines up between the right whales and the smaller rorquals. The gray whale is classified separately—in its own family. The broad flippers are pointed at the tips. When a gray whale cracks the surface, the dorsal hump and a series of bumps, or knuckles, forming the dorsal ridge, become visible. Brownish gray to light gray in color, it is notable for the patches of white and orange that cover its body, some representing barnacles, some whale lice.

The first gray whale champion was Carl Hubbs. In the 1940s, Hubbs moved from the middle of America to the Scripps Institution of Oceanography in San Diego. He was a fish man who became interested in the big grays he saw traveling along the coast. His annual census charted the return of a species from endangered to, more recently, recovered status. The census was revived in the late 1970s by the American Cetacean Society and the Cabrillo Whalewatch Program and operated from Long Point or Point Vicente in southern California. From December 1 to late May, the project invites the public to help identify and make observations about the passing gray whale parade, taking note of behavior and any human impacts.

Thanks to its inshore habitats and migrations along the Pacific west coast, the gray whale has probably led to more people getting interested in watching and studying whales than any other species. Jim Darling started studying grays in the

Previous spread: A humpback calf nudges close to its mother in the warm waters off Tonga. Weaning may begin at only five to six months of age, but most calves don't separate from their mothers until they are at least a year old.

early 1970s after he'd observed them through the summer months off Tofino, British Columbia.

As Darling, who was a student at the time, learned, not all gray whales make the long migration all the way to the Arctic to feed. After spending the winter in the Mexican breeding and calving lagoons, these half-hearted migrators would swim to closer feeding grounds off northern Mexico, California, Oregon, Washington State or British Columbia. Primarily younger animals would stop off to feed in the shallows along the migration route during the summer, staying sometimes until the migrators returned months later, heading back to Mexico.

The eastern North Pacific gray whale is not the only gray whale population. About 100 gray whales spend the summer off Sakhalin Island in the western North Pacific. Some cross the Pacific to visit the eastern North Pacific gray whale lagoons of Mexico, but it is likely that there is a

A mammal-eating killer whale corrals a young gray whale—one of the hazards for gray whales migrating through areas patrolled by killer whales along the west coast of North America.

distinct western Pacific population, a remnant of a population from the pre-whaling era. Where this small population breeds is unknown, although they may migrate to highly degraded habitats off China.

The gray whale was also historically found in the North Atlantic Ocean. Smithsonian Institution researchers in the United States have recovered skeletons of Atlantic gray whales from the outer coast of North Carolina as well as in other Atlantic locations. The Atlantic gray whale went extinct by the 1700s. It was probably already in decline then, and sporadic whaling might have hastened its disappearance from Atlantic waters, but there is little evidence that whaling played a big role in its extinction.

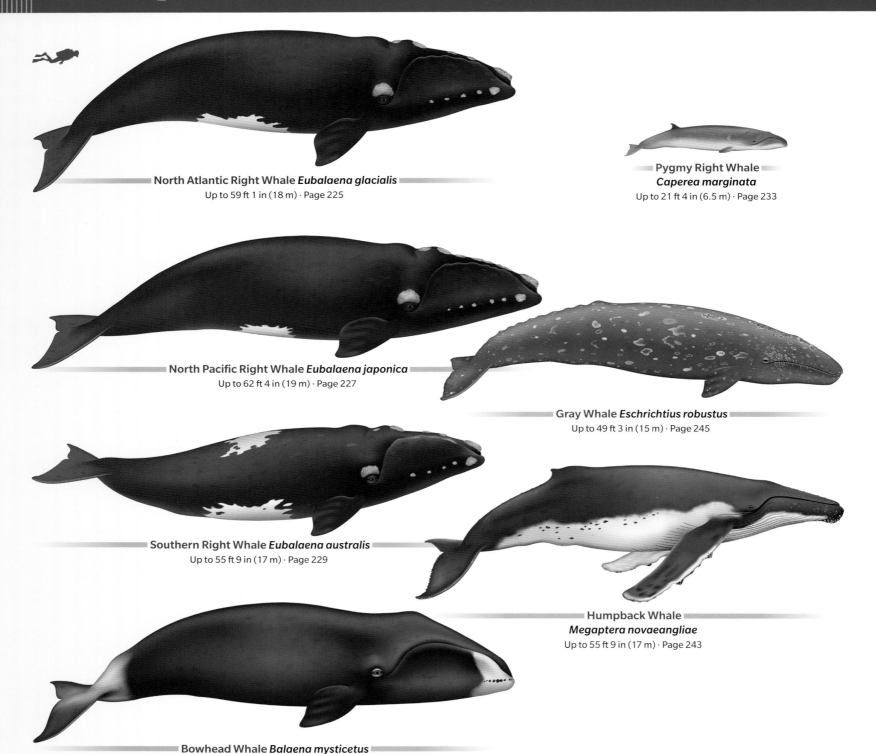

North Atlantic Right Whale *Eubalaena glacialis*
Up to 59 ft 1 in (18 m) · Page 225

Pygmy Right Whale
Caperea marginata
Up to 21 ft 4 in (6.5 m) · Page 233

North Pacific Right Whale *Eubalaena japonica*
Up to 62 ft 4 in (19 m) · Page 227

Gray Whale *Eschrichtius robustus*
Up to 49 ft 3 in (15 m) · Page 245

Southern Right Whale *Eubalaena australis*
Up to 55 ft 9 in (17 m) · Page 229

Humpback Whale
Megaptera novaeangliae
Up to 55 ft 9 in (17 m) · Page 243

Bowhead Whale *Balaena mysticetus*
Up to 65 ft 7 in (20 m) · Page 231

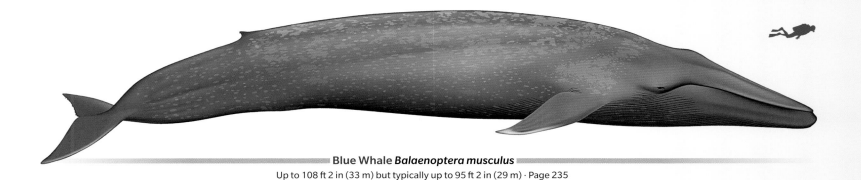

Blue Whale *Balaenoptera musculus*
Up to 108 ft 2 in (33 m) but typically up to 95 ft 2 in (29 m) · Page 235

Fin Whale *Balaenoptera physalus*
Up to 88 ft 7 in (27 m) · Page 236

Sei Whale *Balaenoptera borealis*
Up to 59 ft 1 in (18 m) · Page 237

Common Minke Whale
Balaenoptera acutorostrata
Up to 28 ft 10 in (8.8 m) · Page 241

Antarctic Minke Whale
Balaenoptera bonaerensis
Up to 35 ft 1 in (10.7 m) · Page 240

Bryde's Whale *Balaenoptera edeni*
Up to 54 ft 2 in (16.5 m) · Page 238

Omura's Whale *Balaenoptera omurai*
Up to 39 ft 5 in (12 m) · Page 239

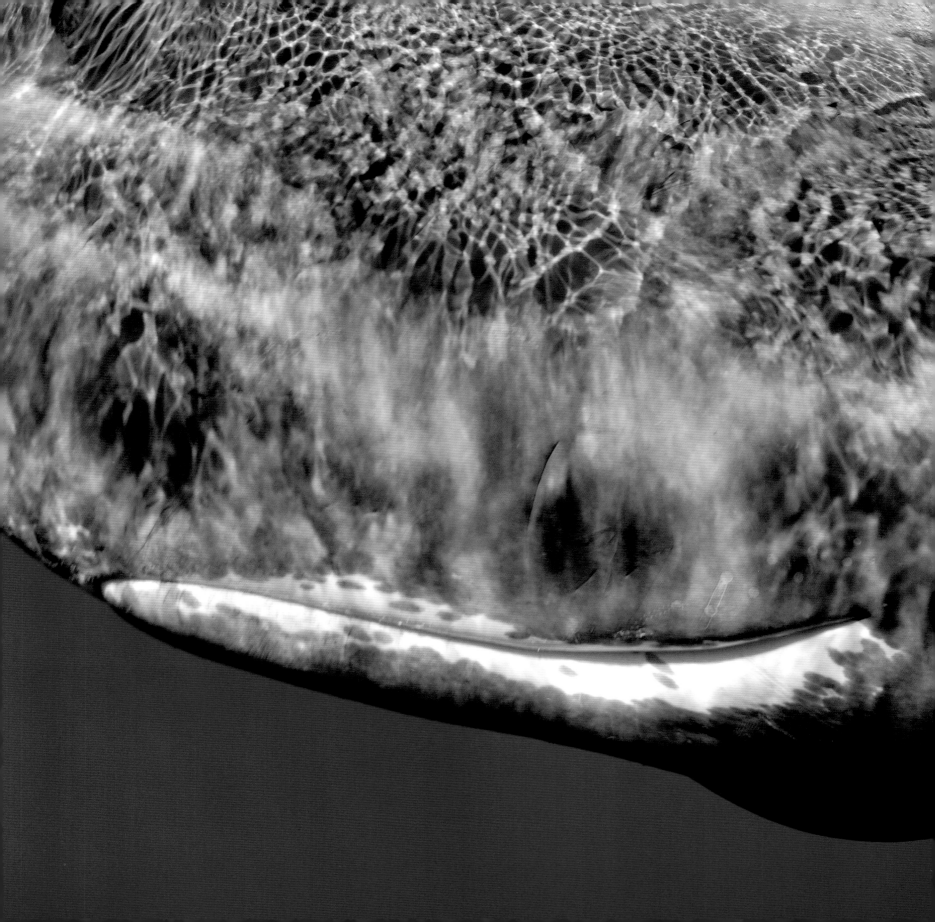

5

Toothed
Whales

Introducing the Toothed Whales: Odontoceti

Large-sized Toothed Whales

Sperm Whale · Family Physeteridae

Sperm Whale

Physeter macrocephalus

The largest of the toothed whales, the sperm whale earned international renown when Herman Melville published his opus *Moby-Dick; or, The Whale*, profiling a malevolent sperm whale and featuring extensive background on the whaling of sperm whales in the North Pacific. To this day, popular images of whales often depict some variation of the sperm whale silhouette, with its massive head, underslung lower jaw and short, wide flippers.

The most famous and widely distributed of all the large, or so-called "great whales," the sperm whale first made its name with American whalers beginning in the mid-1700s, when its oil was used to light the lamps of the 13 original colonies and additional states and territories as the country expanded across the continent to California. Later, Yankee whalers brought their hunting techniques to the Pacific and beyond, teaching the world how to hunt sperm whales and render the oil. The whaling of sperm whales was finally stopped in 1985, after their numbers had been reduced to about one third of their original, from approximately 1 million to 330,000 today. Only Japan has kept on whaling sperm whales commercially, although at a low level. The traditional hunting community of Lamalera, Indonesia, also continues to catch sperm whales most years.

Biologist Hal Whitehead first encountered sperm whales on a sailing expedition through the Indian Ocean in 1982, after the designation of the Indian Ocean Sanctuary by the International Whaling Commission. Spending day after day with extended family groups near Sri Lanka, Whitehead and colleague Jonathan Gordon filmed, photographed and recorded the whales. They were impressed by the whales' shyness and their intense protectiveness toward the calves and juveniles. In a maneuver called the marguerite formation, the whales arranged themselves defensively around the vulnerable members of the group, facing inward toward the center of a circle.

Previous spread: A female sperm whale goes into her dive. The open-mouth posture of one of the sperm whales swimming off Dominica, facing page, could be a sign of protectiveness toward a calf.

A male sperm whale lifts his flukes in the waters off New Zealand.

Alternatively, sperm whales might bunch up and face an interloper, heads up. Bite marks from killer whales and sharks can be seen on many sperm whale flukes, but sperm whales are not often subdued in killer whale attacks.

The fundamental social unit of sperm whales contains roughly 10 females and calves. These groups stay largely in tropical and warm temperate areas, some of them living in the same general area year after year. Large senior males, often living alone in high latitudes, travel across and even between ocean basins, visiting the mother-calf groups from time to time for mating purposes. They usually spend only a few hours with a group before moving on. Sperm whale society is polygynous, with males mating with multiple females.

Before they reach breeding age in their late 20s, the bachelor bulls travel together—too large to be involved with the mother-calf-juvenile groups yet not in a position to command mating privileges.

In their study of whale culture, Hal Whitehead and Luke Rendell paint a picture of how sperm whales may have become so successful as a species. The key, they suggest, was the evolution of the sperm whale's unique spermaceti organ, the sperm whale nose. The huge nose, which makes up most of the head, occupies from one quarter to one third of the whale's body length. The nose contains a click-producing mechanism, and these clicks are focused and formed through the fine oil and nasal sacs connecting air passages before they are sent out to "read" the sperm whale's environment. Whitehead and Rendell argue that this is the "natural world's most powerful sonar system,"

and it has helped the sperm whale to dominate the deep middle layers of the ocean, which is teeming with squid.

The main competition in these deep dark waters comes from other sperm whales. This means that as a species, sperm whales emphasize long-term care of their young and the learning of social skills and cooperation. These social groups are thought to form partly because calves and juveniles, unable to dive into the depths to catch squid, must be looked after. Thus, the females, in effect, organize calf care, taking turns at diving for around 40 minutes, sometimes up to more than an hour, while other members of the social unit stay near the surface with the young. Working their network of relationships, sperm whales feed and take care of each other. This transfer of information is the essence of whale culture.

Pygmy Sperm Whale and Dwarf Sperm Whale · Family Kogiidae

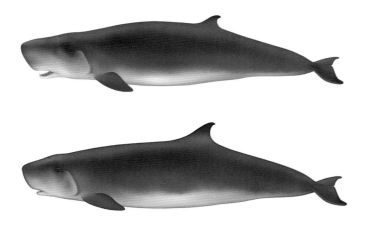

Two other whales carrying the sperm whale common name are the **pygmy sperm whale (*Kogia breviceps*)**, top illustration, and the **dwarf**

sperm whale (*Kogia sima*), bottom. They belong to a different family, and they are much smaller, more dolphin-sized, yet they look vaguely like a cross between a miniature sperm whale and a kind of shark.

Both species have a false gill slit formed from a line on the side of the head and a dorsal fin, unlike the much larger sperm whale with its dorsal ridge. Their teeth are sharp and somewhat sharklike and are found in the lower jaw, fitting into sockets in the upper jaw as with sperm whales. Some pygmy and dwarf sperm whales have a few pairs of teeth in the upper jaw. Both species live in offshore tropical to warm temperate waters of the world ocean.

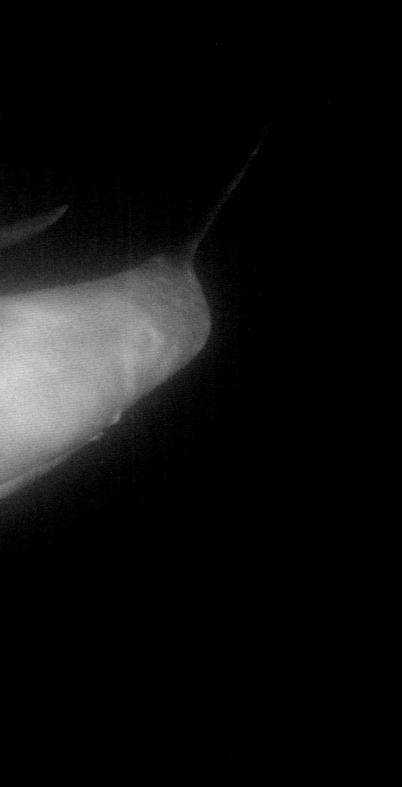

Beluga

Delphinapterus leucas

Much has been said about the dolphin smile. A dolphin can open and close its mouth, but the expression is fixed. Not so with the beluga, whose rubbery face is capable of many expressions. The forehead, or melon, can expand or contract, and the short beak has fat lips that move. The entire head, in fact, can be turned from side to side. However, as supple as the beluga is, it carries a blubber layer up to six inches (15 cm) thick. Biologists use the word "robust" to describe the mature beluga, but belugas are chubby, and some have thick rolls and folds of fat all along their bodies.

A beluga calf more closely resembles a narwhal calf than it does its own mother. The calves of both species are fairly slender and gray to brownish gray in color. The beluga calf turns white between five and 12 years of age, fattening up in the process. The beluga has teeth in the upper and lower jaws suitable for handling a wide range of prey that

A beluga mother and her offspring enter the Churchill River near Churchill, Manitoba. The belugas return to productive river mouths every summer to feed in the open, ice-free waters.

Above: The bright sun flashes off the backs of a group of beluga whales feeding in the mouth of the Churchill River, Hudson Bay. Inset: A newborn beluga calf nuzzles its mother—note its characteristic dark purplish coloring and fetal folds.

varies by area. In general, this whale sticks to shallower areas for feeding, although it can dive deep for food when needed. In summer, belugas come by the thousands to gather in Arctic estuaries, sometimes swimming upriver. They are grouped often by age and sex, some all male and some mixed female/young groups, but membership in these groups fluctuates.

The beluga is rated Near Threatened on the IUCN Red List. There are at least 29 recognized "stocks," or population units, some of which are threatened by environmental contamination—for example, the St. Lawrence belugas—even though hunting has ceased. Other Arctic beluga populations are still hunted. Large numbers in the Russian Okhotsk Sea are being captured for aquaria with deaths occurring during capture. Like the bowhead and the narwhal, the ultimate fate of the beluga is wrapped up in the development of the Arctic and the loss of prey and other potential impacts from climate change. Arguably, its estuary-loving habits put it in greater danger than the other two Arctic whales.

Narwhal

Monodon monoceros

With a sturdy body and a small, rounded head, the adult narwhal is almost white in color, with dark Dalmatian-like mottling. The mature male reaches a length of up to 15 feet 9 inches (4.8 m) and weights of over 3,500 pounds (1,600 kg). It lacks a dorsal fin, but its most attention-getting characteristic is its single protruding tooth, or tusk, a male sexual trait.

In the long summer days in the eastern Canadian Arctic, mature male narwhals like to cross tusks with each other, raising them high out of the water while making "a strange, sad whistle." University of Washington narwhal researcher Kristin Laidre says that there is often a female between the two males. What is going on? Laidre says that sword fighting, breaking ice, spearing fish or digging in the seafloor are popular but mistaken explanations.

"The tusk cannot serve a critical function for narwhals' survival," says Laidre, "because females, which do not have tusks, still manage

An all-male group of narwhals swims and feeds together in the deep offshore waters of the Canadian Arctic.

The tusk of the adult male narwhal is a tooth that penetrates the left-hand side of the upper jaw, growing to nearly 9 feet (2.7 m).

to live longer than males and occur in the same areas while additionally being responsible for reproduction and calf rearing." Instead, as Charles Darwin originally suggested, the narwhal tusk is an example of sexual selection. The tusk acts as a sexual trait similar to the antlers of male mountain goats or deer or the fancy plumage seen in male birds. Perhaps the males with the longest tusks have higher social rank when it comes to mating with a female; crossing tusks may be a way to measure and compare them.

One of two teeth, the tusk of the male narwhal penetrates the left-hand side of the upper jaw and grows out, spiraling counterclockwise, to a maximum length of nearly 9 feet (2.7 m). Sometimes, a right-hand tooth erupts as well, though it does not grow as long. The females are toothless and tuskless, normally, but in rare cases, a female also grows a tusk. There is no evidence that the tusked females or the two-tusked males gain sexual favors or other advantages from their unusual deformity.

Of all whales, the narwhal is the most remote denizen of the High Arctic, even more so than the bowhead whale or the narwhal's fellow family member, the beluga. The bowhead and beluga both have some populations located farther south in cold temperate waters. The narwhal spends the dark winters feeding in the dense Arctic pack ice, diving deep and sucking up choice Greenland halibut and occasionally small squid. It uses echolocation to hunt and catch fish as well as to navigate to fissures in the ice that provide breathing holes.

The narwhal is a champion deep diver, built for endurance. Like a runner training for an annual marathon, the narwhal undertakes only shallow dives in summer when it stays close to land, gradually building up as it migrates toward the deep waters of the wintering grounds. Using satellite tags, Laidre found that over the six-month Arctic winter, a narwhal dives down to at least 2,600 feet (800 m) some 18 to 25 times a day. In addition, it spends more than three hours per day below 2,600 feet (800 m), with half of the dives reaching at least 4,900 feet (1500 m) and lasting around 25 minutes. The record dive depth is more than a mile deep—approximately 5,900 feet (1,800 m).

At 4,900 feet (1,500 m), pressures are intense—more than 2200 PSI (150 atmospheres), and there is no oxygen. For a diving mammal to achieve these depths, it needs to be streamlined, with a flexible, compressible rib cage, and it needs to carry a lot of oxygen-binding myoglobin in its muscles. Researchers have calculated that an average-sized narwhal can carry more than 18 gallons (70 l) of oxygen in its muscles, blood and lungs. Shutting off blood flow to less critical body parts and swimming at a speed of about 1 meter per second, provides enough oxygen for 20 plus minutes underwater.

For centuries, Inuit and other Arctic peoples have hunted the narwhal for its valuable tusk as well as for the meat. In the coming decades, with the melting of Arctic sea ice and ecosystem-related climate-change effects, the narwhal's world may turn upside down. As more humans and animal species enter the Arctic, narwhal food supplies of fish, shrimp and squid may be degraded. With increased ship traffic, there will be more noise, oil spills and contaminants entering the ecosystem. Killer whales that sometimes hunt these whales may become more successful. The narwhal has faced climatic changes before but never at this fast pace and in combination with human pressures from all sides. At present, the narwhal is rated Near Threatened on the IUCN Red List—a recognition that care needs to be taken with this extraordinary, mythic whale. Of note, climate change was not taken into account when this Near Threatened rating was assigned. The acceleration of climate change means that revision is needed soon.

The Social Life of Baird's Beaked Whales

The high cliff above a rugged camp at Poludennaya Bay, on western Bering Island, in the Russian Commander Islands, is the perfect place to sight one of the rarest whales in the world. When the rough seas calm down and the fog lifts, researcher Ivan Fedutin, from Moscow State University and the Kamchatka Branch of the Pacific Institute of Geography, climbs the nearby hill with binoculars. Often he sees humpback whales; once in a while, he sees killer whales; and once in a great while, he sees Baird's beaked whales. When he does, Fedutin and his fellow researchers race down to the water, jump into the boat and head out to sea.

Out on the water, the Baird's beaked whales spout, revealing their long, scratched torsos at the surface, before performing the usual beaked whale stunt of disappearing for up to half an hour and sometimes for more than an hour. Baird's belong to the beaked whale group comprising 22 deep-diving species that like to suck up squid in the deep trenches, often far from land. In the remote Commander Islands, however, deep waters lie close to shore.

Whalers still hunt Baird's beaked whales off northern Japan, despite minimal understanding of beaked whale behavioral ecology. This pioneer Russian study has discovered that Baird's have enduring, stable associations, with some whales preferring the company of certain other whales. These are so-called fission-fusion societies in which whales come together for a time and then split up, although some stay with the same whale buddies over a period of at least six years.

More than 145 whales have been individually identified and at least 20 have been determined to be seasonally resident to the area.

This Russian Cetacean Habitat Project focuses on humpback whale and killer whale studies too. It works closely with the Kamchatka-based Far East Russia Orca Project, which has a catalog of more than 2,000 individual killer whales. Both of these long-term projects were founded and are supported by Whale and Dolphin Conservation.

Baird's beaked whale society differs from the matrilineal groups found in killer whales, which are formed around the mother and stay together for life. Of all the beaked whales, the Baird's beaked whale is the largest at up to 36 feet 5 inches (11.1 m) long for females and 35 feet 1 inch (10.7 m) for males.

The Baird's is also turning out to be among the most social of the beaked whales, with an average group size of eight individuals. Photographs display the scarred torsos used for identification—all Baird's beaked whales have scars largely made by the other whales they associate with. The individuals most heavily scarred are the older whales in their 40s, 50s or even 80s.

Some of the scars, however, come from killer whale teeth. About 15 percent carry such scars, which suggests that killer whales try but don't always succeed in hunting these whales. The larger group size and close associations in Baird's beaked whales may therefore be partly explained as an anti-predator tactic. Not only are larger groups probably less likely to be attacked by orcas, but if they are attacked, Baird's beaked whales can work together to care for their young and defend themselves, and there is less probability that an individual whale might be taken.

Other scars come from suspected cookie cutter sharks, indicating long-range travels into warmer waters, perhaps during the winter months. Still other scars come from close calls with drift nets, the telltale net marks leaving a crisscrossed imprint. These deadly "walls of death" can

kill anything and everything that swims or bumps into them. Often left for days at a time, a typical net in the western North Pacific may catch prized salmon but can also scoop up sharks, seabirds, turtles, porpoises and whales. A few animals manage to get away with scars; others manage to break away with parts of the net wrapped around their jaws, flippers or tail flukes, which can mean a slow death over many years. Most suffocate quite quickly, unable to extricate themselves.

Baird's beaked whales were once hunted in small numbers around the North Pacific. The last whaling of these whales persists off Japan. About 50 to 65 are killed every year, with unknown impact on the population. Since 1987, more than 1,000 Baird's beaked whales have been killed in Japanese waters.

For the most part, conservation groups and governments in the North Pacific region have yet to consider this whale worthy of "saving." Now that researchers are learning more about the social lives of the Baird's beaked whale, however, the species is taking on a character all its own.

Baird's beaked whales are the subjects of continuing behavioral ecology research in the waters off Bering Island, in the Commander Islands State Biosphere Reserve. Researchers from the Russian Cetacean Habitat Project, with support from Whale and Dolphin Conservation and Moscow State University, have identified individual whales through marks on their backs and found that these whales have long-term associations.

The Whale that Swam into London

On January 20, 2006, there were reports that a whale had been spotted swimming up the River Thames in England. British Divers Marine Life Rescue identified the species as a **northern bottlenose whale (*Hyperoodon ampullatus*)**, and crowds gathered to watch as the whale neared central London. This species had never been recorded in or anywhere near the Thames; in fact, the whale's home waters were in the deep seas north of Scotland, extending up to the Arctic and west to Canada.

The next day, the 19-foot (5.85 m) female whale was gently captured and loaded on to a barge so that she might be returned to the open sea. But as the barge approached the mouth of the Thames, the whale suffered convulsions and died before being released.

The necropsy showed that the juvenile female had suffered severe dehydration and kidney damage. Various scenarios were suggested. The whale may have been lost and took a wrong turn. She may already have been ill before she came into the Thames. She would have had trouble finding her deep fish and squid prey in the shallow North Sea, much less upstream in the Thames.

Home waters for northern bottlenose whales include Iceland's Skjalfandi Bay, where whale watchers have seen them in the midst of the friendly local minke whales. Scottish and then Norwegian whalers hunted them in the 19th century for the spermaceti oil in their big square heads.

The forward portion of the melon on the head of the males is particularly prominent. In older males, the melon is pale and often pushed in or flat, thought to be due to the propensity of northern bottlenose males to head butt each other in breeding squabbles. In the 1960s, Norwegian whalers discovered the offshore canyon east of Nova Scotia called The Gully and killed many of the whales there. In 1988, researchers led by Hal

A rescuer attempts to help a northern bottlenose whale that had become lost in the River Thames in the heart of London, England, in January 2006.

Whitehead from Dalhousie University in Halifax found that The Gully was a productive biologically diverse habitat still being used by some northern bottlenose whales. Decades of detailed study produced much of what we know about this species, as well as the protection of that habitat. The Gully Marine Protected Area is the first protected area focused on a beaked whale species.

The Thames whale left behind several positive legacies. First was the connection that Londoners made with the deep sea. Second, the northern bottlenose whale became almost a household name—she belonged to a species native to British waters mainly off northern Scotland, and people recognized that northern bottlenose whales deserved protected habitat in the deep waters that they call home. Finally, there was an awareness that marine mammals do swim in the Thames, that it has a connection to the open ocean and that there are some direct results of continuing to improve the health of the river. The Zoological Society of London reports that between 2004 and 2014, more than 2,500 marine mammals were seen in the Thames, including 49 whales and 450 dolphins and porpoises. Some of them were in trouble, but many were not. Canary Wharf is said to be the best viewpoint for whale, dolphin and porpoise sightings.

Beaked Whales · Family Ziphiidae

Beaked whales are the most mysterious of all the whales. Small to moderately sized whales, they live mostly far out to sea, diving deep in search of squid, their main prey. Like most toothed whales, beaked whales are social, although the groups tend to be small, containing up to seven individuals. Researchers lucky enough to encounter beaked whales with their low, often faint spouts, might manage a few photographs, but then the whales disappear, plunging down, down, sometimes for an hour or more.

There are 22 beaked whale species, and counting. Through the process of comparing genetic data, often from materials taken from whale strandings on remote beaches or from historic collections, three new species have been named since 1991. These mainly Pacific Ocean beaked whales include the **Perrin's beaked whale (*Mesoplodon perrini*)**, named after the master taxonomist William Perrin, whose definitive list of cetacean species and subspecies is revised annually (Visit: marine-mammalscience.org/species-information/list-marine-mammal-species-subspecies).

Most male beaked whale species have only two teeth in the lower jaw, positioned anywhere from the tip of the beak to midway along the jaw line, while the females have no erupted teeth. The teeth—more properly called tusks—often protrude from the lower jaw in males. Some individuals have multiple barnacles growing on the tusks that can be seen when the whales surface. Two species, the **Baird's beaked whale (*Berardius bairdii*)** and the **Arnoux's beaked whale (*Berardius arnuxii*)**, have two pairs of

A New Member of the Family

In 2016, cetacean geneticists and field researchers led by Phillip Morin from NOAA Southwest Fisheries Center in the United States and others, including the Far East Russia Orca Project, directed by Alexander Burdin and the author, with fieldwork by Olga Filatova and Ivan Fedutin, announced their finding of a rare, new beaked whale. This soon-to-be-declared new species was formerly thought to be a small black form of the Baird's beaked whale. For now, nobody knows the conservation status of this whale. It has most likely been caught in whaling operations, along with the larger Baird's beaked whale. Without genetic research to back the proposal to recognize this as yet unnamed species, it might well have slipped into extinction.

tusks, while the **northern bottlenose whale (*Hyperoodon ampullatus*)** and the **southern bottlenose whale (*H. planifrons*)** have one prominent pair and sometimes a second pair. The **Shepherd's beaked whale (*Tasmacetus shepherdi*)** is the only beaked whale with a full set of functional teeth; both males and females have them, but in addition, the males have two enlarged tusks at the tip of the lower jaw. One species, the male **strap-toothed beaked whale (*Mesoplodon layardii*)**, sports fancy tusks that grow from the sides of the lower jaw up and around the beak, preventing the mouth from opening fully. For all beaked whales, in fact, these tusks have limited use for grabbing, chewing or eating, and the females have no teeth at all. As such, beaked whales probably use suction to seize and ingest their prey.

It is thought that the tusks are used in sparring competitions between males of the same species, resulting in extensive scarring, especially on the older male bodies. The spacing and patterns of parallel scars created by the tusks can sometimes be used to help identify the species.

Most beaked whales live only in one part of the ocean, though information on their distribution is limited and subject to periodic revision. Two exceptions are the **Cuvier's beaked whale (*Ziphius cavirostris*)**, which is found in every ocean ranging close to polar waters, and the

Blainville's beaked whale (*Mesoplodon densirostris*), which has a similar though not as extensive distribution in colder waters. Some beaked whales have Northern and Southern Hemisphere forms. No beaked whales are considered to be threatened, but mostly this is a reflection of how little we know about them: 20 of the 22 species are officially rated Data Deficient by the IUCN Red List. Scientists know only enough about two of the species to be able to give them a Least Concern rating: Cuvier's beaked whale and the southern bottlenose whale.

As researchers extend their studies farther offshore to more remote locales, we've begun learning more about beaked whales. Cuvier's beaked whales, which have shown themselves to be susceptible to navy mid-frequency sonar, have been the subjects of studies in places where they have stranded in the Canary Islands and off the Bahamas in the North Atlantic, as well as in the Mediterranean, where sonar exclusion zones have been proposed to protect their habitat. The northern bottlenose whale became famous when a young female swam up the Thames in 2006, but even before that, this species was the first beaked whale to be studied intensively using individual photo identification.

In 2004, the Canadian government created a marine protected area around a large submarine canyon called The Gully, located 124 miles (200 km) off Halifax, Nova Scotia. Some 16 species of whale are found here, including the northern bottlenose whale, subject of long-term photo-ID research. Working since 2007, the Russian Cetacean Habitat Project, sponsored by Whale and Dolphin Conservation, has found long-term associations among some of the more than 145 photo-identified individual Baird's beaked whales

A deep-diving Cuvier's beaked whale performs a rare breach in the Ligurian Sea waters of the Pelagos Sanctuary for Mediterranean Marine Mammals, protected in 2000 by Italy, France and Monaco.

around Bering Island, Russia. More recently, Blainville's beaked whales have been the subjects of individual photo-ID studies in the Bahamas. Diane Claridge's pioneer work has revealed the existence of harems made up of an adult male living in a group of up to seven whales with multiple adult females. The male and his harem appear to get the best feeding spot, while groups of other adults and subadults are left to search for food in less productive areas.

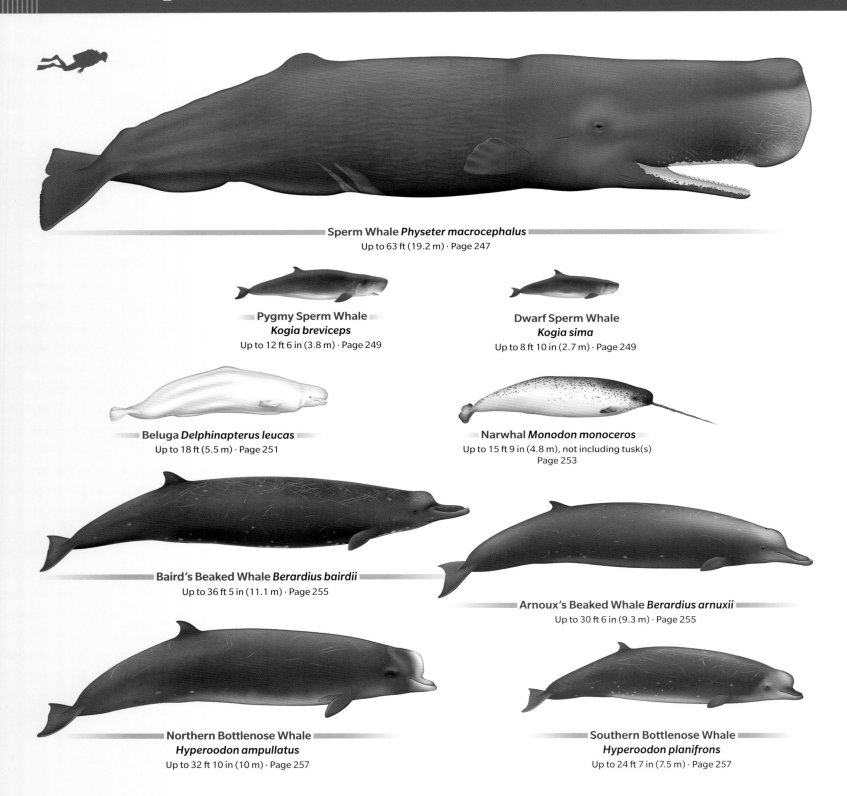

Sperm Whale *Physeter macrocephalus*
Up to 63 ft (19.2 m) · Page 247

Pygmy Sperm Whale
Kogia breviceps
Up to 12 ft 6 in (3.8 m) · Page 249

Dwarf Sperm Whale
Kogia sima
Up to 8 ft 10 in (2.7 m) · Page 249

Beluga *Delphinapterus leucas*
Up to 18 ft (5.5 m) · Page 251

Narwhal *Monodon monoceros*
Up to 15 ft 9 in (4.8 m), not including tusk(s)
Page 253

Baird's Beaked Whale *Berardius bairdii*
Up to 36 ft 5 in (11.1 m) · Page 255

Arnoux's Beaked Whale *Berardius arnuxii*
Up to 30 ft 6 in (9.3 m) · Page 255

Northern Bottlenose Whale
Hyperoodon ampullatus
Up to 32 ft 10 in (10 m) · Page 257

Southern Bottlenose Whale
Hyperoodon planifrons
Up to 24 ft 7 in (7.5 m) · Page 257

Longman's Beaked Whale
Indopacetus pacificus
Females up to 21 ft 4 in (6.5 m) · Page 258

Sowerby's Beaked Whale
Mesoplodon bidens
Up to 18 ft (5.5 m) · Page 258

Andrews' Beaked Whale
Mesoplodon bowdoini
Up to 14 ft 5 in (4.4 m) · Page 259

Hubbs' Beaked Whale
Mesoplodon carlhubbsi
Up to 17 ft 9 in (5.4 m) · Page 259

Gervais' Beaked Whale
Mesoplodon europaeus
Up to 15 ft 9 in (4.8 m) · Page 260

Ginkgo-Toothed Beaked Whale
Mesoplodon ginkgodens
Up to 17 ft 5 in (5.3 m) · Page 260

Gray's Beaked Whale
Mesoplodon grayi
Up to 18 ft 5 in (5.6 m) · Page 261

Hector's Beaked Whale
Mesoplodon hectori
Up to 14 ft 1 in (4.3 m) · Page 261

Deraniyagala's Beaked Whale
Mesoplodon hotaula
Up to 15 ft 9 in (4.8 m) · Page 262

Strap-toothed Beaked Whale
Mesoplodon layardii
Up to 20 ft 4 in (6.2 m) · Page 262

True's Beaked Whale
Mesoplodon mirus
Up to 17 ft 9 in (5.4 m) · Page 263

Perrin's Beaked Whale
Mesoplodon perrini
Up to 14 ft 5 in (4.4 m) · Page 263

Pygmy Beaked Whale
Mesoplodon peruvianus
Up to 12 ft 10 in (3.9 m) · Page 264

Stejneger's Beaked Whale
Mesoplodon stejnegeri
Up to 18 ft 8 in (5.7 m) · Page 264

Spade-toothed Beaked Whale
Mesoplodon traversii
Up to 17 ft 5 in (5.3 m) · Page 265

Blainville's Beaked Whale
Mesoplodon densirostris
Up to 15 ft 5 in (4.7 m) · Page 266

Shepherd's Beaked Whale
Tasmacetus shepherdi
Up to 23 ft (7 m) · Page 267

Cuvier's Beaked Whale
Ziphius cavirostris
Up to 23 ft (7 m) · Page 267

6

Dolphins and Porpoises

Medium-sized Toothed Whales

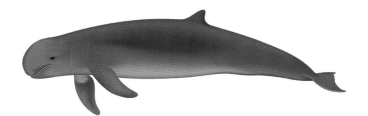

Irrawaddy Dolphin
Orcaella brevirostris

In the fierce heat and humidity of the Sundarbans mangrove in Bangladesh, the sight of a group of Irrawaddy dolphins seems as strange as it is a welcome reprieve from the weather. The Irrawaddy dolphin looks a bit like a beluga calf—it is gray to off-white with a low-lying dorsal fin and a blunt-nosed face that shows a distinct similarity to the Arctic-loving beluga, even though the nearest belugas are 3,500 miles (5,630 km) away. Also, like the beluga, this dolphin has mobile lips that can spit water; it has been known to do so while feeding, which may help it capture fish.

Bangladeshi researchers Elisabeth and Rubaiyat Mansur and Brian D. Smith from the Wildlife Conservation Society work from a local nature tourism boat owned by Rubaiyat's father. When they started work on the local dolphins in 2002, they had no idea what they would find. To their surprise, they discovered that the dolphins were living in an area of high cetacean diversity with seven different species native to Bangladesh's rivers, estuary, coast and offshore waters. One of these, the Irrawaddy dolphin, turned out to have the healthiest population numbers of that species in the world. Elsewhere in their range, wherever they have been studied, Irrawaddy dolphin populations number no more than in the low hundreds, with some down to only a few tens of individuals. In Bangladesh waters, the numbers have been estimated at about 5,800 individuals.

The Irrawaddy dolphin is not classified as a "river dolphin," and yet it is found in the estuaries, rivers and lagoons of South and Southeast Asia and adjacent islands. In Myanmar, they live nearly 900 miles (1,450 km) up the Ayeyarwady River (also known as the Irrawaddy River). They also penetrate deep into the rainforests along the Mahakam River of Borneo and into the Mekong of Cambodia and Lao PDR, as well as into Chilika Lake in India and Songkhla Lake in Thailand. These five freshwater populations are all considered Critically Endangered by the IUCN Red List. Overall, the Irrawaddy dolphin is rated Vulnerable.

Many dolphin generations ago, when these oceanic dolphins first moved into the rivers, they would have found good fishing opportunities and could thereby avoid competing with the various other coastal or oceanic dolphin species along the shores of these countries. Over the past few decades, fishing nets have been taking too many fish and also catching dolphins accidentally—sometimes they become entangled and die in nets, as so-called bycatch. These rivers, estuaries and lagoons have also become more polluted

and crowded with boat traffic that threatens the dolphins' health.

For now, the Irrawaddy dolphin population in Bangladesh is relatively healthy. The model research and conservation work of the Wildlife Conservation Society Bangladesh Program has led to habitat protection in Bangladesh waters, and this will help to ensure a future for dolphins here.

Previous spread: A pair of common bottlenose dolphins takes to the air. Above, a tropical Irrawaddy dolphin displays the extraordinary neck flexibility that allows this species to turn its head around and even to see what is behind it.

Discovering the Dolphins and Porpoises

The group of species that we know as "dolphins" are as confusing to the public as they are to scientists. When most people say "dolphin," they mean the bottlenose dolphin, the species with a worldwide distribution that is most often seen in aquaria and in the wild. In fact, there are two species of bottlenose dolphins, the common (*Tursiops truncatus*) and the Indo-Pacific (*Tursiops aduncus*), and, in all, there are five families of dolphins comprising 48 total species. Let's sort it out.

The public confusion starts with the dolphin vs. porpoise debate.

Scientists are sure about this one. There are 37 species of oceanic dolphins with features that include, among other things, conical teeth and a generally larger size than porpoises. There are only seven species of porpoises, and they have spade-shaped teeth. Yet in some regions of the world, the word dolphin and porpoise are used interchangeably, and in some areas, the word porpoise is used to indicate any small cetacean, that is, any dolphin or porpoise.

Next, the complete list of dolphins includes eight species whose names feature the word "whale." We can start with the killer whale,

or orca. Taxonomically the largest dolphin, the orca seems decidedly undolphinlike to the public, not least because it sometimes eats dolphins and porpoises as well as large whales. There are also the false killer whale, the pygmy killer whale, the melon-headed whale and the short-finned and long-finned pilot whales. All of these are dolphins. And don't forget the northern right whale dolphin and the southern right whale dolphin. With both whale and dolphin in the name, what can you expect? These two species are in fact neither whale nor dolphinlike. They were called "right whale dolphins"

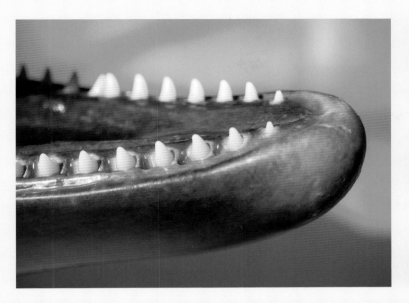

Facing page: The common bottlenose dolphin, left, and the harbor porpoise, right, show the differing head shapes of dolphins and porpoises. Above, the conical teeth of a bottlenose dolphin, left, and the spade-shaped teeth of a Dall's porpoise, right.

because, like right whales, they have no dorsal fin. Yet they are tiny compared with right whales. They are elegant, smooth-backed, black and white dolphins with no relationship to right whales.

Then there are the river dolphins. The true river dolphins comprise four families of the currently recognized four species, including the extinct baiji, or Yangtze River, dolphin. However, one of these river dolphins, the franciscana, spends most of its time in estuaries and along the coast and not in rivers. In addition, several oceanic dolphins, the Delphinidae family, spend some or even all of their time in rivers. They are not taxonomically river dolphins, but their habits, their lives, are sometimes riverine.

These include the Irrawaddy dolphin from south and southeast Asia, the tucuxi and Guiana dolphins from eastern South America, and several others. The tuxuci and Irrawaddy dolphins can be found hundreds and even thousands of miles from the open sea. Even the bottlenose dolphin sometimes adapts to feeding and living in an estuary or lower river system. And there is one porpoise— the narrow-ridged finless porpoise— that sometimes lives far up rivers, including the Yangtze in China.

The confusion to scientists comes with taxonomic uncertainties. Many dolphins have intermediate forms. A species may look somewhat like one species but also carry characteristics of another. Sometimes genetic

studies suggest splitting species, when the appearance of the individuals does not seem to justify it. Other species, such as the bottlenose dolphin and the killer whale, have multiple forms, or ecotypes, that seem to demand that they be split into several more species. Taxonomy takes time and careful comparison of museum and lab materials that may be scattered around the world. Even with detailed field studies that draw on genetics, a species is, to paraphrase Charles Darwin, what a body of experts decides is a species, based on the best available information. The problem is that it is expensive and time-consuming to convene experts to address every taxonomic question.

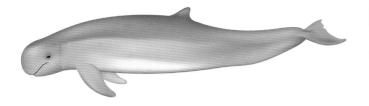

Australian Snubfin Dolphin

Orcaella heinsohni

The Australian snubfin dolphin is the common name for a newly created species, split off from the Irrawaddy dolphin. It looks much like the Irrawaddy dolphin but has a darkish back, lighter sides and a whitish stomach. It lives mainly along tropical northern coastal Australia where it travels in groups of fewer than 10 individuals.

The snubfin dolphin can be identified as individuals from marks and scars on the low-lying dorsal fin and back. These scars are reportedly from encounters with the larger, more robust Australian humpback dolphins that chase and aggressively interact with the snubfin dolphin. Sometimes mating also occurs between the two species, and a number of hybrids have been witnessed. They are found in fairly low numbers, on the order of 100 to 200 individuals, in the few areas where they've been studied.

Blackfish Dolphins

"Blackfish dolphins" is not a formal grouping but the name given to six of the largest dolphin species, all with "whale" in their name, all predominantly black in color. The name blackfish is still used by artisanal fishers and mariners in the Caribbean, among other places around the world.

The six dolphin members of the blackfish club are the killer whale, or orca, the short-finned pilot whale, the long-finned pilot whale, the false killer whale, the pygmy killer whale and the melon-headed whale.

The **killer whale (*Orcinus orca*)** is the top predator in the sea—unmistakable in the typical pod formation with the characteristic tall dorsal fin of the males, up to 6 feet 7 inches (2 m) high, which is unique among all whales and dolphins.

The **short-finned pilot whale (*Globicephala macrorhynchus*)** and the **long-finned pilot whale (*G. melas*)** are in the same genus—*Globicephala*, which means globelike head, a fitting description for these whales, also sometimes called potheads or pothead whales. The two species are nearly impossible to distinguish in the areas where they overlap at sea. The number of teeth and length of the flippers varies in each species, but these characteristics are difficult to recognize in the field. The long-finned pilot whale prefers cold temperate waters—these are the whales driven to shore and killed every year by the Faroe Islanders in the North Atlantic. The short-finned pilot whale prefers warm and tropical waters. These whales, too, are hunted off Indonesia and in the Caribbean, and they help support a large whale-watching industry in the Canary Islands.

The **false killer whale (*Pseudorca crassidens*)** is found in mainly coastal tropical to temperate waters and gets its name from the much larger, more robust black and white killer whale. The false killer whale is a slender, black-bodied dolphin, highly social with some stable associations and site fidelity, based on field studies in Hawaii and Costa Rica. These whales often mix with other smaller dolphin species, but they are also known to act aggressively, attacking other dolphins and even large whales. They mainly eat fish, including wahoo, mahi mahi and tunas, as well as cephalopods.

The largely tropical **pygmy killer whale (*Feresa attenuata*)** is rarely encountered or studied. Highly social like other blackfish, the population numbers, however, appear to be small. It usually moves slowly compared with other blackfish and shows strong site fidelity, at least in Hawaii.

Finally, the **melon-headed whale (*Peponocephala electra*)** is a tropical dolphin that overlaps the pygmy killer whale distribution and is often confused with that species. The melon-headed whale, however, swims faster and has a much larger group size of 100 up to 500 individuals. Melon-headed whales love to socialize with other dolphins. Through photo identification, they have been shown to have site fidelity around Hawaii.

Killer Whale, or Orca

Orcinus orca

Coastal residents and researchers in far-flung communities around the world are getting to know killer whale communities almost as an extension of their own human communities.

• In late 2016, the ancient matriarch Granny, also known as J2, was seen for the last time with her extended pod, the J pod. People on the shores of Puget Sound and the Salish Sea off British Columbia and Washington State cheered her on. J2 then disappeared and was presumed dead.

• Off California, members of the "L.A. pod" made short work of a great white shark while local researchers and whale watchers stood by. Following the kill, the sharks disappeared from the area for weeks. The L.A. pod was thought to have moved south and wasn't seen again for several years.

• In Iceland, researchers working closely with whale-watching operators have discovered that certain well-known killer whales feed on summer-spawning herring during the winter, then move away to check out the food off Shetland in northern Scotland, 700 miles (1,130 km) away.

The foundation of our knowledge about killer whale communities around the world is the photo-identification method pioneered by Michael A. Bigg and his colleagues in the 1970s off Vancouver Island, Canada. They are able to identify individual orcas through photographs of the small nicks and notches that appear on the trailing edge of the dorsal fin, the overall shape of the fin and the shape and markings on the saddle patches. This allows researchers to establish the relationships between individuals, to track their movements and resightings, to catalog their vocalizations and genetic information and to assign new calves to known mothers. Bigg and his team were the first researchers to identify the two main killer whale ecotypes in the northeast Pacific: fish-eating (resident) and mammal-eating (transient), now referred to as Bigg's killer whales. In the North Atlantic, some killer whales are believed to alternate between fish and seals. In the Antarctic, there may be four different ecotypes. Type A feeds on minke whales. Type B hunts seals in the ice. Type C feeds on certain fish, and the little-known Type D may eat the Patagonian tooth-fish. Worldwide, there are at least 10 ecotypes, but with the many orca studies now in every ocean, more ecotypes may emerge in the future.

One such long-term study, which I co-direct with the Russian marine mammal researcher Alexander Burdin and team leaders Olga Filatova, Ivan Fedutin and Tatiana Ivkovich, has been based in the Russian Far East since 1999 and in the Commander Islands since 2007. The Far East Russia Orca Project (FEROP) sponsored by Whale and Dolphin Conservation, Humane Society International, Animal Welfare Institute and others, has found two killer whale ecotypes: fish-eating (resident) and mammal-eating (transient), the same ecotypes as are found in the northeast Pacific off the west coast of Canada and the United States. FEROP researchers have observed resident killer whales feeding on fish (salmon, Atka mackerel, cod) and transient killer whales feeding on marine mammals (minke whales, fur seals, Dall's porpoises). The fish-eating type forms matrilineal groups of up to 10 or more individuals and travels in pods (seven to 20 or more individuals) and communities of various related pods with up to several hundred individuals. Mammal-eating orcas travel in smaller, more fluid groups, usually two to six individuals. In the northeast Pacific, in addition to these two ecotypes, researchers led by John K. B. Ford and Graeme Ellis have also found a shark-eating killer whale ecotype called "offshores," which are not known from Russian waters.

Fish- and mammal-eating orcas in the North Pacific have small differences in the appearance of their dorsal fin and saddle patch. Genetic analysis has shown that killer whale ecotypes belong to reproductively isolated populations, with different mitochondrial DNA haplotypes. Acoustically, the killer whale calls of mammal-eating killer whales have significantly lower frequencies than those of the fish-eating killer whales. This is true on both sides of the North Pacific and the transient frequencies are also lower than North Atlantic killer whale populations.

This large male killer whale from the southern resident orca community lives off southern Vancouver Island, in Puget Sound, the inland Salish Seas, and along the outer coast of the northwest United States.

A female killer whale approaches the surface to exhale and breathe.

Avacha Gulf on the southeastern coast of the Kamchatka Peninsula is a critical habitat for killer whales. Foraging takes up about half of the killer whale activity budget in the area. From 2000 to 2005, Atka mackerel was one of the main prey for killer whales. By 2006, this productive Atka mackerel spawning ground in Avacha Gulf was depleted due to local overfishing. Salmon became the main killer whale prey in the Gulf.

Besides foraging, killer whales use Avacha Gulf for resting, socializing and giving birth. The large multi-pod aggregations in killer whales do not lead to cooperative foraging. Such aggregations are thought to be "clubs" in which whales gather to establish and maintain social bonds; they appear to play a role in reproduction.

The Russian killer whale photo-ID catalog includes more than 700 killer whales in Avacha Gulf and off eastern Kamchatka and about 1,100 killer whales around the Commander Islands. For the entire Russian Far East, the total number of photo IDs stands at more than 2,000 fish-eating and 130 transient mammal-eating orcas. According to the identification rate, most of the killer whales using Avacha Gulf, the main study area in eastern Kamchatka, are already identified. New calves have been found in recent field seasons, but very few new adults have been found. In the Commander Islands, however, the level of identification of new whales is still high. This suggests there is a greater number of individuals in this area and that many of them are visitors rather than residents.

Some mixing of killer whale pods occurs between Avacha Gulf, the Commander Islands and other regions of Eastern Kamchatka, but the amount of mixing varies for different areas. A high rate of matches was found between the coastal areas of Eastern Kamchatka (from Avacha Gulf to Karaginsky Gulf)—apparently, killer whales in these areas belong to one community. The number of matches between the Commander Islands and Avacha Gulf is low. In the Commander Islands, FEROP researchers Filatova and Fedutin have encountered many whales that were not seen anywhere else. This suggests that a significant number of killer whales come to the Commander Islands from some unknown areas (most likely, the Aleutian Islands, in U.S. waters).

In general, we have found that the Commander Islands are the crossroads for killer whales from different regions, which explains the large number of identified animals but the low repeated

Killer whales live in tight family groups in a matrilineal society. The males stay with their mothers for life. Female calves, once they reproduce, form their own units but continue to remain close to their mother.

sighting or recapture rate. Avacha Gulf, on the other hand, is the core area and an important feeding area—a critical habitat for the local killer whale community during the summer months.

The appearance and structure of the Russian fish-eating killer whale communities resemble the well-known northern and southern Vancouver Island communities and the resident Alaskan communities—but there is no known exchange between any of them. Genetic studies show that the resident Avacha Gulf killer whale haplotype

All-white Whales and Dolphins

Melville's Moby-Dick was a fictional white sperm whale, yet even today, white whales and dolphins of various species are sometimes seen at sea. They have included Risso's and bottlenose dolphins and southern right, humpback, killer and sperm whales, among others. In recent years, whale watchers and scientists have come to know and love Migaloo, the all-white humpback whale that has made frequent visits to eastern Australia. Belugas, also known as white whales, are naturally white, yet they start out life a bluish gray color, only becoming all white when they mature.

In the Russian Pacific waters, around the Commander Islands, our research team, the Far East Russia Orca Project (FEROP), has encountered three all-white killer whales in repeated sightings that began in 2008. We saw them at various times in unrelated pods. One is a probable female named Mama Tanya, another a calf named Lemon, and the last is a mature bull called Iceberg. In 2014, in the Okhotsk Sea, three additional white killer whales appeared, a female accompanied by a juvenile and another juvenile. Other reports of white orcas in the Northwest Pacific bring the total up to at least five and

Iceberg, a white male killer whale, is at least 22 years old. A probable albino, he is in apparently good health, living with his fish-eating pod of about 13 individuals, all of whom have normal pigmentation.

possibly eight white individuals.

In August 2010, we met the big male we named Iceberg—a sighting reported around the world. Iceberg is notable for his creamy yellow coloring, which can look bright white in some light conditions. He has only the faintest signs of the killer whale eye and saddle patches. A mature bull, he travels in a pod of about 12 individuals. In 2010, he was traveling alongside two other bulls, possibly his

siblings, and following behind either a young male or a mature female that might have been his mother or sister. All of the other whales in the pod were typical-looking fish-eating orcas.

Albinism is caused by defects of melanin production that affect the whole skin and eyes, while leucism is due to defects in pigment cells and tends to be patchy. Iceberg's creamy yellow color suggests that he is albino rather than leucistic. Five to eight white killer whales in a population in the low thousands could be a result of inbreeding in the past. It is known that killer whale populations are relatively small with low genetic diversity. In some mammals, albinistic animals may be more susceptible to infection or other disease and may stand out to predators. Of course, with killer whales, predation is not a problem.

The only white killer whale definitively confirmed as albino was a young female named Chimo, captured from a transient eastern North Pacific pod off southern Vancouver Island in 1970. She lived for two years at Sealand of the Pacific in Victoria, British Columbia, before dying of a rare genetic disorder, Chediak-Higashi syndrome, which is thought to have caused her albinism and left her vulnerable to infections.

Our FEROP team has returned to the Commander Islands every summer since 2010, searching for Iceberg and his pod. On July 14, 2015, off the Commander Islands, FEROP researchers met him again. He was in his family group surrounded by other pods numbering more than 100 orcas. This new sighting puts Iceberg's age at 22 years at least. Iceberg has survived into adulthood, and it is clear that his albinistic pigmentation has not compromised his health.

is the same as that of the southern killer whale community and of some Alaskan and Aleutian residents. These killer whales share a more recent common ancestor, while the northern killer whale community of fish-eating residents and the transient marine-mammal eaters are more distantly related.

The continued work and presence of our FEROP team in the Russian Far East is valuable not only for the science it produces but also because it allows us to monitor populations and solve conservation problems. Since 2012, an industry has developed in the Russian Far East in which killer whales are caught and sold to aquaria in Russia and China. At least 16 killer whales were taken between 2012 and 2015. The killer whales were captured in the western Okhotsk Sea and are all thought to be transient mammal-eating killer whales. There are no current estimates of the numbers of transient killer whales in Russian waters. Russian marine-mammal scientists, as well as the International Whaling Commission and various conservation groups, have expressed concerns about the growing number of quotas and permits that are given to capture killer whales without reliable abundance estimates. FEROP and other such science projects that focus on conservation biology have a stake in the future of whales. No outcome is guaranteed. The success or failure of conservation efforts for killer whales depends on local researchers, communities, tour companies and supportive government wildlife managers in these and other countries around the world. Good science is needed, as well as a precautionary approach and respect for what we don't know. There also has to be a willingness of government to listen and to implement conservation work.

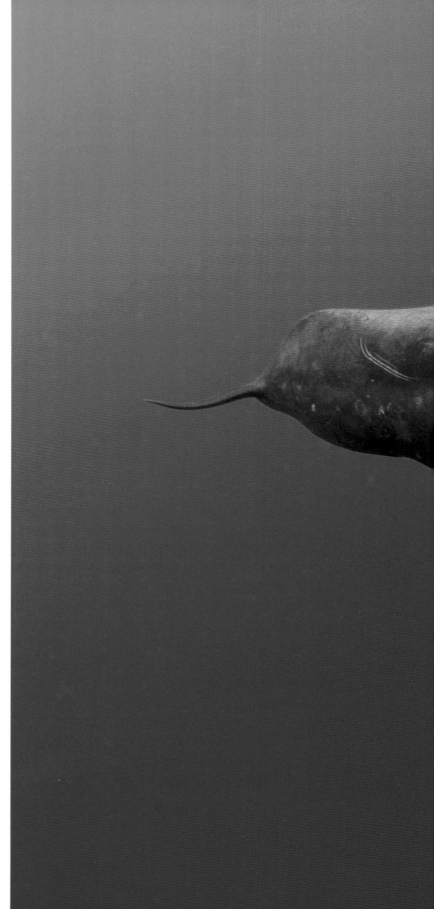

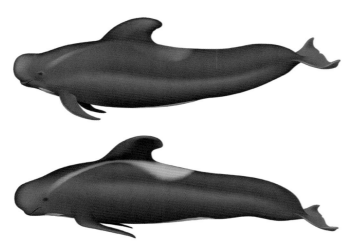

Short-finned Pilot Whale

Globicephala macrorhynchus

Long-finned Pilot Whale

G. melas

At the northern edge of the North Atlantic around the Faroe Islands between Scotland and Iceland, local people use their powerboats to drive a huge group of passing dark gray long-finned pilot whales onto the beach. The ritual of the *grindadráp*, or grind, is about 1,000 years old, with an average of 838 pilot whales and 75 dolphins killed every year (1709–2011). Because of the social nature of pilot whales and other dolphins, the grind carries none of the risks of whaling or other dangerous work. Once the whales are in the shallow waters, the hunters bring out their knives and try to sever the spinal cords. Even if it were

possible for some whales to escape, they would not leave their podmates.

The bulbous-headed pilot whale is among the most social of all cetaceans, and humans have long exploited its highly social nature. In the past, it has been hunted in large groups in eastern Canada and the United States, around Japan, Greenland, in the Caribbean, as well as in the Faroe Islands hunt, which continues even today. When these numbers are added to those that strand, it becomes clear that sizable numbers of pilot whales die on beaches year after year in various parts of the world.

Efforts to return pilot whales to the sea once they've stranded usually fail. They may be disoriented but still healthy, yet they refuse to go. Being social is a valuable trait in many ways, in that it can help protect a whale and her close relatives from predators. Young group members can also learn from their more experienced elders.

But it is also worth pondering—as has the whale biologist Toshio Kasuya, who discovered menopause in pilot whales—that if a social trait continues to be relentlessly exploited by humans, and environmental conditions regularly lead to mass deaths, whether perhaps that part of whale culture will eventually be diminished.

Previous page: The highly social short-finned pilot whale lives in large, close-knit groups in tropical and subtropical waters. Mass strandings of this species occur frequently, in part because if one whale is ill, disoriented or gets stranded on a beach, the group will not leave—to the point of becoming stranded themselves.

Risso's Dolphin
Grampus griseus

R isso's dolphins are the only light-colored large dolphins. With their ghostly form and large dorsal fin, they are clearly identifiable, even from a distance. Close-up, the scars and scratches cover their bodies, especially on the older individuals.

The coloring, or pigmentation, of Risso's dolphins varies by location. Tropical Risso's have medium to dark gray backs, while those in cooler temperate waters are splotchy grayish white with black eye patches and a prominent dorsal fin. The usual Risso's diet is thought to be squid, which means

The bodies of Risso's dolphins are heavily scratched. These marks are especially noticeable on males, which turn whiter as they age.

they generally stay offshore in deeper waters than do many fish-eating dolphins. They have teeth only in the lower jaw, which suggests that they may suck up the squid just as beaked whales do.

The scratches and extensive scarring in older Risso's dolphins implies that they have many stories to tell. Scars can be a result of aggression, competition for mates and/or play, but we are only beginning to learn the Risso's dolphin's story.

Few dolphin study areas compare to the stark, moody beauty of the Hebrides. With its diversity of marine species, it is Scotland's national marine treasure.

Off the northeast Isle of Lewis, Whale and Dolphin Conservation (WDC) researchers Nicola Hodgins and Sarah Dolman wait patiently for the weather to change and the dolphins to arrive. In Scotland, it's possible to study minke whales, harbor porpoises and bottlenose dolphins and to have frequent sightings on most good-weather days. Hodgins and Dolman, however, have taken the more difficult route. They're studying one of the lesser-known dolphins, the Risso's dolphin, in one of the more remote corners of Scotland.

But in August 2015, for three days in a row, they had Risso's dolphins in abundance.

Day one: At least eight Risso's dolphins puff into view. Although standoffish, the dolphins are within telephoto range. We manage to secure their photo IDs. All eight turn out to be new to the study, adding 10 percent new fin prints to our photo-ID catalog.

Day two: Seven Risso's show up, friendly, sociable and familiar—all of them having been photo-identified together in previous years. Later, we meet short-beaked common dolphins, including one strangely pigmented individual, possibly a hybrid with another dolphin species.

Day three: With the seas up and weather changing, Risso's appear again. This time only three individuals can be photo-identified. One individual is from the previous day. In the midst of the group, we spy a lone common dolphin—it is either another suspected hybrid or the one from the previous day looking for friends.

On the River with Guiana Dolphins

As we glided along the Suriname River outside the city of Paramaribo, in Suriname, we were so low in the water that we could touch the river. First one dolphin then another swam over, turning, splashing and belly flopping beside us. We were at eye level, staring at the slender pink-bellied **Guiana dolphin (*Sotalia guianensis*)**.

The dolphins moved around in small groups of four to eight individuals, with the calves mixed in with the mothers and other adults. They spyhopped in a way I've seen before with some bottlenose dolphins, bobbing their head two or three times, watching us in the boat before going under. Some of them jumped and performed a half twist, turning their bodies before they landed.

The pink-bellied slivers of spirited fun stood out against the muddy river and the green jungle all around. In the tangle of the mangrove, we could see caimans. The bright scarlet ibis,

Suriname's national bird, took wing along the shore, painting bits of fire against the green forest and blue sky. We were in the land of sloths and anteaters, one of the most untouched pieces of rainforest left in South America—a fact not lost on conservationist Monique Pool. Pool had organized the trip to explore the possibility of protecting this part of the river and the estuary for the dolphins and the terrestrial wildlife that she and her citizen science team have been studying for the past decade.

Later, the researchers showed me their hundreds of dolphin photographs—the nicks, notches and scratches on the fins and back provide plenty of detail that will allow them to get numbers, track movements and chart associations.

Then we met up for a full-day workshop with the representatives of the country—ecologists, conservation groups, army, navy, mayor, environment and tourism ministries, as well as Pool and her group—to discuss the

The Guiana dolphin was only recently separated from the tucuxi, shown at left. Both live off tropical and subtropical eastern Latin America, but the tucuxi often stays well upstream in rivers, while the Guiana dolphin is coastal and estuarine, rarely venturing upstream.

possibility of developing the dolphin tourism in a healthy way and making a marine protected area. The result was a multi-year plan with a goal to integrate the science and management with the neighboring countries of French Guiana and Brazil.

Guiana dolphins are not numerous compared with other dolphin species, but they do have a future if communities like those in Suriname come together to cherish and support research and conservation efforts. The main threats appear to be bycatch and pollution in their riverine habitat. Guiana dolphins are rarely hunted, yet researchers have reported that products derived from Guiana dolphins were being sold as amulets or love charms in the Brazilian Amazon while others have been used as bait for the shark fishery.

The Guiana dolphin was only recently separated from the **tucuxi (*Sotalia fluviatilis*)**, the closely related freshwater *Sotalia* species that lives far up the Amazon River and its tributaries as far as Ecuador, southeast Colombia and southern Peru. Smaller than the Guiana dolphin, the tucuxi travel typically in groups of two to four individuals but occasionally up to 30, staying in the main channels of the river, unlike that more famous resident, the Amazon River dolphin, or boto, which likes to spread out into the forest during the flood season.

Ever since starting their study in 2010, Hodgins and Dolman have been seeing hybrids. They believe that the frisky Risso's males are mating with offshore bottlenose dolphins, resulting in hybrid individuals. Sometimes the Risso's may be seen interacting with common dolphins too.

Hybrids of various whale and dolphin species, for example, blue whales and fin whales, are known in other areas of the world, but they are relatively rare. Western Scotland seems to be something of a hotbed for hybrid dolphins.

The minimum number of Risso's dolphins recorded off western Scotland during the summer of 2015 was 68 individuals, many of them resightings, indicating a resident population. Farther down the west coast of Britain, around Bardsey Island, off northwest Wales, another WDC research team has photo-identified what is considered likely to be a separate population of Risso's dolphins. There, a total of 267 individual Risso's have been photo-identified over the past decade, with only a few resighted from year to year.

These and other Risso's photo-ID studies, such as those conducted around the Isle of Man, in the Azores and in the Mediterranean, among other areas, are starting to put a few pieces of the puzzle in place. The research suggests that feeding, nursing, site fidelity and long-term associations occur among at least some individuals. All of the U.K. areas need protection, and there are plans to expand the research and to set up exchanges to compare the various catalogs and to look for matches.

Rough-toothed Dolphin

Steno bredanensis

U nlikely ever to win a dolphin beauty contest, the primitive, reptilian-looking rough-toothed dolphin has a cone-shaped head and a splotchy gray torso tinged with white and pink. Many have circular scars, left behind by cookie-cutter shark bites.

These dolphins sometimes travel with pilot whales and bottlenose or other dolphins. At a distance, researchers might see only the dorsal fin, but from closer range, rough-toothed dolphins stand out. They live in the deep waters across the tropical and subtropical zones of the world ocean. They feed on cephalopods and large fishes, and they may be mahi mahi (dorado) specialists. Typical group size is up to 20 individuals, with occasional aggregations of more than 100.

The pink Indo-Pacific humpback dolphin, right, familiar to Hong Kong ferry travelers and other boaters, executes a half-breach. Illustrations, facing page, show the four different species of the highly coastal humpback dolphins. The Atlantic and Indo-Pacific humpback dolphins are well known to researchers. The Indian Ocean and the Australian humpback dolphins, however, have only more recently been recognized as separate species.

T he humpback dolphins are named for the fleshy raised platform upon which sits a relatively small, swept-back dorsal fin. This is the dolphin version of the humpback whale's dorsal fin/platform arrangement, although few notice the humpback whale's platform when a big female waves her 16-foot (5 m) flippers and then lifts her broad tail high out of the water. The humpback dolphin's platform is sometimes missed, too, and some individuals appear to have no platform at all.

The four species of humpback dolphins are mainly coastal and inshore tropical species, with some ranging into subtropical waters. The gray-colored **Atlantic humpback dolphin (*Sousa teuszii*)** lives along the west coast of Africa. In Mauritania, some of these dolphins join up with bottlenose dolphins to herd mullet toward shore-based traditional fishers in a long-standing annual cooperation between dolphins and local people.

With the **Indo-Pacific humpback dolphin**

Clockwise, from top left: Atlantic humpback dolphin (*Sousa teuszii*), Indo-Pacific humpback dolphin (*S. chinensis*), Indian Ocean humpback dolphin (*S. plumbea*) and Australian humpback dolphin (*S. sahulensis*)

(S. chinensis), the first thing you see is the adult's white to bright pink bodies, which are especially striking when mothers travel beside their dark gray calves. The famous pink dolphins of Hong Kong are celebrated by locals and authorities, yet there is a danger they may be displaced by Hong Kong industrial developments, which include airport expansion and the Hong Kong-Zhuhai-Macau Bridge across the Pearl River Delta. For now, the Hong Kong dolphins are struggling to hold on in the face of noise, ship traffic and an increasingly polluted environment. The airport and bridge developments and the proximity of the dolphins to Hong Kong has meant that more is known about them than about any other of the humpback dolphin species.

The **Indian Ocean humpback dolphin (S. plumbea)** is light to brownish or slate gray and is found in a continuous band from around Cape Town, South Africa, to Malaysia, including the Red Sea and the Persian Gulf. South African photo-ID studies have revealed a fluid social structure, but the members of dolphin groups in Mozambique appear to have more long-standing relationships.

In South Africa, the protected bays are favored for socializing and resting. Mating and calving occurs year-round with a peak from January to March. Life expectancy can be 40 years or more.

The gray-colored **Australian humpback dolphin (S. sahulensis)** sometimes has white areas on the small dorsal fin and only a hint of the humpback platform. This dolphin is famous for using sponges in play, but it may also use them as tools, as has been found with bottlenose dolphins in western Australia. The sponges may be used by the dolphin to protect its beak when it probes the sand for food or to avoid stingray barbs or the sharp spines of catfish, both known prey items.

All humpback dolphins live close to large cities and busy coastal communities in Africa, Asia, the Indonesian archipelago and northern Australia, often near estuaries or other fresh water sources. As a result, they are often caught in fishing nets and other gear and threatened by boat traffic and polluted waters. The future for these dolphins depends on coastal waters being kept clean with well-managed local fishing industries and strategic protected areas.

Common Bottlenose Dolphin

Tursiops truncatus

Bottlenose dolphins were originally considered one species, but in the past few decades, with genetic and other studies, they have been separated into the **common bottlenose dolphin (*Tursiops truncatus*)** and the **Indo-Pacific bottlenose dolphin (*T. aduncus*)**. In the North Atlantic, there is an offshore and a coastal form that may well be separated in future. Additional species may be recognized at some point as well.

Wildlife photographer and dolphin field officer Charlie Phillips from Whale and Dolphin Conservation has a special relationship with the local bottlenose dolphins along the northeast Scotland coast. In January 2015, he photographed a mother dolphin named Kesslet with her four-month-old calf near the Kessock Channel, which crosses their home waters of the Moray Firth Special Area of Conservation (SAC). He'd first seen Kesslet with the baby the previous September. However, since then, Kesslet had turned up alone, the calf having disappeared, probably dead.

A play day for common bottlenose dolphins in the waters of the Moray Firth, Scotland.

In April 2015, Phillips noted in his blog that it "is a difficult time for our dolphins at the moment as the early salmon run has pretty much stuttered to a halt. The dolphins are positioning themselves anywhere that they think migratory fish will run ... they are even coming through the area on a falling tide having a look for fish, which is fairly unusual." Mainly, however, Phillips was worried about Kesslet.

Day after day, when Charlie Phillips went looking, Kesslet was nowhere to be found. The morning of May 1, however, he finally caught up with her near the Inverness Marina. She was hunting for salmon and looked healthy—especially with her prize food catch of the day. Kesslet had successfully raised another calf, one called Charlie, named in Phillips' honor by Aberdeen University researchers. Kesslet's Charlie is now mostly on his own and trying to fit in with the boys. In time, Kesslet might have the chance to be a mother again.

This northernmost population of bottlenose dolphins has some of the largest, most robust dolphins in the world. At 10 to 13 feet long (3–4 m) and 1,100 to 1,300 pounds (500–600 kg) for the adults, they can be twice as long and several times the weight of bottlenose dolphins in other areas. They are centered on the Moray Firth but range along the northeast Scotland coast as far south as the Firth of Forth and beyond to the coast of northeast England.

The bottlenose dolphin is beloved all over the world. Since photo-ID studies began in the 1970s, it has become well known in the wild, with hundreds of populations, or separate breeding groups, living in subtropical to temperate waters of every continent except Antarctica. In most of these locations, a core group lives close to shore, typically around a productive estuary, firth, harbor or inlet, and sometimes near population centers.

Throughout Europe, bottlenose dolphin populations reside in a number of SACs, designated under the European Union's Habitats Directive. In Portugal, colorful sculptures to the local dolphins have been mounted along the Sado Estuary. Along the coast of South Carolina and Georgia, bottlenose dolphins work together to chase mullet onto the beach, then roll up on the shore, high and dry, to grab their catch before rolling back into the water. In Monkey Mia, western Australia, a core group of dolphins come into the shallows to interact with visitors.

The bottlenose dolphin appears to be one of the most adaptable marine mammals, adjusting its habits to fit local ecological conditions and to catch particular prey. Besides living close to shore and near human population centers, separate populations reside offshore where the fishing may require deeper dives and living in rougher seas, with more exposed conditions.

Bottlenose dolphins are also known to be "bully boys." In Scotland, Wales and in California around Monterey Bay and San Francisco, they harass and sometimes kill harbor porpoises. In southeastern Brazil, they have been reported acting aggressively towards Guiana dolphins.

In northeastern Scotland, thousands of people watch their favorite dolphins from land-based sites at Chanonry Point overlooking the Moray Firth, near the North Sea. It is the best spot to find dolphins in the U.K. Other prime sighting areas include the Kessock Channel near Inverness, the mouth of the Cromarty Firth and Spey Bay. These are also good sites to find harbor porpoises and—a little farther out to sea—minke whales.

Below, a common bottlenose dolphin tosses his salmon prize to make it easier to swallow. Common bottlenose dolphins, facing page, show off what is often termed the classic "dolphin smile."

Stenella dolphins show how difficult it can be to sort out dolphin species, not only when they are observed at sea but for taxonomists working with skeletal materials. In the 1960s, researchers learned that large numbers of dolphins were being killed in tuna fishing operations in the Eastern Tropical Pacific. Before that, tuna was captured by pole and line. Following the introduction of nylon monofilament seine nets, fishermen began chasing dolphins instead. The tuna boats would set their nets around the dolphin schools, which often traveled above the tuna, and the dolphins were thus caught too. Often, they suffocated and died.

Conservation groups began taking action to stop the practice. Sam LaBudde's undercover video aboard a tuna boat revealed the suffering of the dolphins and led to the successful "dolphin-friendly" tuna branding campaign. Successful conservation also depended upon scientists determining which species were being killed and understanding the impact on each population. At that time, the taxonomy of *Stenella* species being caught was unclear.

Dolphin scientist and taxonomist William F. Perrin spent years sorting them out. Several *Stenella* species turned out to have separate populations and forms, and it became clear that certain portions of the populations of two species were the main ones getting hit. Chief among them was the northeast Pacific offshore population of the **pantropical spotted dolphin (*Stenella attenuata*)**. Between 1959 and 1972, an estimated 3 million pantropical spotted dolphins died in tuna nets, reducing them to 19 percent of their original numbers. As well, hundreds of thousands of eastern spinner dolphins and common dolphins were killed. These numbers would have been enough to drive some dolphin species to extinction. By the 1980s, the number of dolphins killed was down to 5,000 per year, still 5,000 too many, but a great improvement. But as of 2015, the Eastern Tropical Pacific *Stenella* populations particularly targeted by the tuna fishers had not returned to their former abundance.

Most *Stenella* species are widely distributed throughout the world. They are also caught as part of bycatch off Thailand and in other areas, as well as during dolphin-hunting activities in the Caribbean, Sri Lanka, Indonesia, the Philippines and Japan.

Stenellas prefer warm water, hanging out mainly in tropical and subtropical seas. Two species, the **striped dolphin (*S. coeruleoalba*)** and **Atlantic spotted dolphin (*S. frontalis*)**, also range into cooler waters. Along with the Atlantic spotted dolphin, the **clymene dolphin (*S. clymene*)** is restricted to the Atlantic Ocean, including Caribbean and Gulf of Mexico waters. The striped dolphin was known to the ancients, and their images can be found on the walls of ruins in Crete. The **spinner dolphin (*S. longirostris*)** is profiled on the following spread.

Previous page: Pantropical spotted dolphins glide through the waters off Baja California. Atlantic spotted dolphins in the clear waters of the Bahamas, facing page, have been the subject of underwater behavioral studies for several decades.

Spinner Dolphin
Stenella longirostris

One of the best-studied dolphins, the spinner dolphin is a well-loved wildlife feature of island atolls and tropical and subtropical bays in every ocean. Other dolphins may occasionally spin with a couple of revolutions, but the spinner's spin is unique, and its characteristic aerial spinning behavior makes it instantly identifiable. It starts its spinning underwater, just before emerging with a head-up breach that clears the water surface. By the time it reaches the leap's full 10-foot (3 m) height, it may rotate the long axis of its body several times. Seven spin rotations is the recorded maximum. Once one spinner has started spinning, others will join in, and not just with one spin per dolphin but with many, in what can seem like a frantic spinning session.

Why do these dolphins spin? Multiple motivations are suspected. Spinning may be a social display related to courtship, or it may be a practical exercise intended to remove remoras or other parasites. It may also be a dolphin's way of flexing muscles and getting the blood flowing after naptime.

The daily lifestyle of the spinner is well organized around meals. Most spinner dolphin groups hunt at night, using their sonar to catch a wide variety of midwater fishes, as well as shrimps and squid, many of which undertake nocturnal vertical migrations up the water column. After feeding all night offshore, groups of 50 or more come into their favorite sandy, shallow bays in the morning and rest until late afternoon or evening. As they prepare for the nighttime feed, they take to the air, spinning, and this can be combined with tail and flipper slaps and racing about.

Spinner dolphins vary in appearance with at least five recognized geographic forms, several of which are considered subspecies. The Central American, eastern and whitebelly spinner dolphins are only found in the Eastern Tropical Pacific, while the dwarf spinner dolphin lives in southeast

In the evening, above left, spinner dolphins charge offshore to spend the night hunting and feeding on fish that ascend the water column. Above right, a spinner dolphin practices his craft, spinning around up to seven times before falling back into the sea. Spinning gets the group excited—once one spinner starts spinning, others follow.

Asia, Indonesia and northern Australia. The most common form of spinner, sometimes referred to as Gray's spinner dolphin, is found throughout the ocean except in limited parts of tropical Asia and the Eastern Tropical Pacific. Besides the complexity of forms, the social structure also varies considerably. Some populations have fluid arrangements, the so-called fission-fusion societies in which dolphin groups periodically join up and then separate, re-forming into different associations each time. Other spinners are known to live for many years in stable groups.

Spinners have become tourist attractions in Hawaii, the Ogasawara Islands of Japan, Fernando de Noronha off Brazil and in the Red Sea, especially around spinner resting areas. In places where they can be watched from shore, there is less chance of disruption to the sleepers. Dolphin researchers argue that careful regulations are needed to control any interactions with resting spinners as these can disrupt natural behavior with consequences for their ability to reproduce.

Short-beaked common dolphins (*Delphinus delphis*), as well as the closely related **long-beaked common dolphins (*D. capensis*)**, are resident to warmer oceanic waters. Each has a tall dorsal fin, slender flippers and distinctive markings. Both have pronounced beaks, with one species having a slightly longer beak.

Common dolphins can travel in groups of 400 to 500 individuals. To view such a mass movement of life and energy is to witness one of the ocean's great wildlife productions. Off South Africa, huge groups of common dolphins hunt sardines in the midst of feeding Bryde's whales, seabirds, sea lions and sharks. Underwater video shows the dolphins deftly herding the fish and then grabbing them as if they were on a platter.

In the eastern Mediterranean, in the waters off Greece, there might be at most 20 to 25 individuals, according to photo-identification studies. Yet common dolphin sightings have recently increased and may indicate a modest return. Even so, their survival will depend on the health of their fish prey as well as their ability to avoid fatal interactions with fishing gear. Mediterranean researchers have identified important dolphin habitat, but Greece has yet to take steps to protect it.

Some people contend that the Mediterranean provides a prediction of the future of the world ocean. With 30 percent of the ship traffic on 1 percent of the ocean, the Mediterranean is more

In the warm waters of the Savu Sea in eastern Indonesia, a Fraser's dolphin, facing page, finesses an elegant leap.

heavily polluted and more intensely fished by the countries that straddle its borders than is any other part of the ocean. Wars and migrant crossings have been a common daily feature of life in states on the southern and eastern borders of the region. Many people are surprised that whales and dolphins still persist in the Mediterranean. Today, the future of that habitat is in question.

Fraser's Dolphin

Lagenodelphis hosei

This tropical and subtropical, mainly deep-water dolphin has a robust torso with a short beak and a relatively small dorsal fin and set of flippers. The Fraser's dolphin was discovered from skeletal materials in 1956, and it wasn't until the 1970s that these dolphins began to be recognized in the wild.

Little is known about their social behavior and ecology. Traveling together in groups of hundreds to thousands, sometimes in mixed cetacean schools, they splash acrobatically and smash against the waves. They are sometimes hunted in Japan and the Caribbean as well as in other countries and are subject to bycatch, but their numbers are thought to be healthy, especially in the Eastern Tropical Pacific.

The subjects are photogenic, but the so-called "Lag" dolphins, the six dolphins from the *Lagenorhynchus* genus, can be a hard group to study. Try taking photo identifications of individuals when a few hundred dolphins are speeding by your boat and the action is fast and furious. Persistent efforts might result in sharp dorsal fin photos, thus making a handful of individuals clearly identifiable. These individuals can become "markers" for the group. In this way, it is possible to estimate numbers and establish site fidelity, but making detailed behavioral notes and sorting out the relationships is more difficult.

Lags are robust dolphins with short, thick beaks. The six lag species include the **dusky dolphin**, **hourglass dolphin**, **Peale's dolphin**, **Pacific white-sided dolphin**, **white-beaked dolphin** and **Atlantic white-sided dolphin**. All lags have stripes of one kind or another along their flanks. Some are thin lines; others are complex patterns of lines and patches.

The **dusky dolphin (*Lagenorhynchus obscurus*)** is the best-studied lag, and there are several papers and a book devoted to its biology and behavior.

The **hourglass dolphin (*L. cruciger*)**, the dolphin typically seen from ships en route to Antarctica, has a jet-black back and lower flanks with a striking white area shaped like an hourglass splashed across each flank.

In appearance, the **Peale's dolphin (*L. australis*)**, found only off southern South America, looks like a cross between the Pacific white-sided dolphin and the dusky dolphin, although only dusky dolphins overlap in part of its range. The Peale's dolphin can be distinguished by the big black stripe that extends along its flank as well as by its dark, masklike face.

The **Pacific white-sided dolphin (*L. obliquidens*)** prefers the warm- to cold-temperate waters of the North Pacific. It has delicate, complex black, gray and white coloring with fine line demarcations and a two-toned black and gray dorsal fin.

Both the **white-beaked dolphin (*L. albirostris*)** and the **Atlantic white-sided dolphin (*L. acutus*)** are seen on whale-watching trips in the northeast United States, Quebec and eastern Canada, as well as off Scotland, Norway and Iceland. The distribution of white-beaked and Atlantic white-sided dolphins overlaps across the cooler waters of the North Atlantic, but the white-beaked ranges into colder waters farther north along the east and west of Greenland and even farther north than Svalbard in the Arctic. As the name suggests, the white-beaked dolphin usually has a light-colored to white beak. The Atlantic white-sided dolphin has a yellowish tan patch on the rear flank. Sometimes, when these dolphins skip between the waves, the tan patches flash in the sun like a signal light.

Pacific white-sided dolphins swim just beneath the surface of the water.

Commerson's Dolphin

Cephalorhynchus commersonii

Biologist Vanesa Tossenberger's long-term study of a small population of dolphins in San Julián, in far southern Argentina, is part of the reason she is International Policy Director for Whale and Dolphin Conservation, working to protect neglected whale and dolphin populations in South America and around the world.

Tossenberger's partner Miguel Iñíguez, Fundación Cethus president, reports that 2015–2016 marked the 20th year of their research. They and their team have amassed 84 photo IDs of the **Commerson's dolphin (*Cephalorhynchus commersonii*)**, and in 2015, there were 24 resighted and four new dolphins, all calves. In addition, they keep track of Commerson's and franciscana dolphins (*Pontoporia blainvillei*) in other Argentine estuaries. Iñíguez notes that unless we look after even the small populations, which are the breeding units, the species themselves will go extinct.

The Commerson's dolphin belongs to the *Cephalorhynchus* genus, which comprises four species, all of which live close to shore in relatively small populations, mainly or entirely in one country and all in the cold temperate Southern Hemisphere. The New Zealand **Hector's dolphin (*C. hectori*)** is Endangered and declining. The **Chilean dolphin (*C. eutropia*)**, mainly resident in Chile and Tierra del Fuego, Argentina, is Near Threatened and declining. The **Heaviside's dolphin (*C. heavisidii*)** of Namibia, western South Africa and Angola is Data Deficient, as is the Commerson's dolphin. The Commerson's dolphin lives mainly in separate populations around river estuaries and bays in Argentina, but some are found in Chile, in the offshore Islas Malvinas (the Falkland Islands) and in the French Kerguelen Islands.

The *Cephalorhynchus* name comes from the Greek *kephale*, or "head," and *rhunklos*, meaning "snout," which suggests their common anatomical characteristic of having a gradual slope from the head to the blunt-nosed snout without a defined melon or beak. The most visible feature, however, is the rounded dorsal fin, said to look like a "Mickey Mouse" ear.

These small dolphins might be taken for porpoises. Reports often declare that they are shy and hard to locate, but experience teaches that they show typical dolphin curiosity toward humans. They will swim figure eights under a boat and leap out of the water, though perhaps not as often or as acrobatically as the larger dolphins. All four have a mixture of black and white coloring, but each species has different patterns. Commerson's have the most dramatic coloring, with a wide band of white that extends across the back to the white underside. When Commerson's dolphins swim their underwater figure eights, they look like black and white optical art—op art in motion.

Dusky Dolphin Society

Working with Roger Payne off southern Argentina in the 1970s, U.S. researchers Bernd and Melany Würsig quickly fell in with the local dolphins and encountered the greatest diversity of dolphin species in the world. This is where Charles Darwin marveled at biological diversity on his five-year voyage with Captain William Fitzroy on the *Beagle*, which eventually led to Darwin's theory of evolution by natural selection. He named one of the local dolphins after his friend and *Beagle* ship captain William Fitzroy. Today, it's known as the dusky dolphin, though the South American subspecies still has Fitzroy's name. In addition to duskies, the closely related endemic Peale's dolphin, the somewhat more southerly cold-water, offshore hourglass dolphin, and the inshore, river-loving endemic Commerson's and Chilean dolphins, with their rounded dorsal fins, can also be found at the southern tip of South America. The bottlenose, Risso's and southern right whale dolphins are also seen here. But thanks to their aerial acrobatics, the duskies command the most attention.

A dusky dolphin leap is a work of art. Up they go out of the water, one, two or eight at a time, several dolphin lengths above the surface, turning and tumbling. Würsig and his students have spent decades studying various dolphin species, but duskies are a favorite—jump for jump, Würsig claims, they are the most active dolphins in the world. Bow-riding or traveling dolphins might leap synchronously as they travel in one direction, but half a dozen or more duskies can launch simultaneously in a small patch of ocean, each heading in a different direction, yet forming a complex aerial ballet.

What does this leaping behavior achieve? Probably many things— backslaps designed to make noise, lengthwise leaps that reduce water resistance, splashless head-first jumps that allow the jumper to scout for feeding birds in the distance, and the highly acrobatic somersault leaps that create a partylike atmosphere. The latter may reflect the behavior of dolphins meeting new dolphins and engaging with old friends. Regardless of its function, leaping involves "high group alertness," with signaling or the inciting of excitement all part of the action.

Ruminating on dusky dolphin society after four decades of research, Würsig remarks that, apart from brief interludes for rest, these dolphins are almost always interacting. These interactions suggest that individuals can recognize other individuals. "But these are fission-fusion societies," says Würsig, a reference to a social order in which the size and composition of the group are perpetually changing.

Imagine the daily challenges of life in a fission-fusion society: The fission aspect means the dolphins split into small subgroups of one to 12 individuals for mating, foraging and performing nursery duties or socializing in mixed age or sex groups. The fission of groups—especially nursery groups—may be necessary to avoid rambunctious males or predator killer whales or sharks. The choice of fission or fusion can also be related to the specific prey, with smaller prey schools or individual prey requiring fission into small subgroups.

Strategies for finding and herding huge prey balls, on the other hand, require fusion into a big group that can search, stay in communication and then combine energies to pin the prey against the surface. This may also explain some of the dusky dolphin's leaping behavior. These dolphins live and travel in groups of 20 to 500 and sometimes up to

more than 1,000 individuals. When the big social groups cooperatively feed on schooling fishes, those high, water-clearing breaches may be the best way of coordinating the group foraging behavior.

All of this endless splitting, joining, rejoining and monitoring relationships that develop may well account for the large brains in these and other dolphins.

Dusky dolphins are famous for their high, complex leaps and group leaping sessions. Researchers suggest dolphins may leap for a number of reasons: as a social activity, to scout for food or just to have fun.

Saving the New Zealand Dolphin

In 2013, alarmed by the declining numbers of these dolphins, Whale and Dolphin Conservation researchers Mike Bossley, Gemma McGrath and I talked with New Zealand dolphin researcher and Professor Elisabeth Slooten to help prepare a map showing the areas that needed protection from the trawls and set nets. Then we engaged economist Tristan Knowles from Economists at Large to conduct a survey of New Zealand citizens to determine how much this dolphin was worth to them and whether protecting more of their habitat would receive support.

A survey of 1,000 New Zealand residents revealed that 80 percent would support strong measures to save Hector's dolphins. Some 52 percent of respondents were willing to pay some annual amount as a dolphin protection tax, while 63 percent were also willing to pay more for fish if dolphin protection measures led to increased consumer prices. At the same time, 57 percent said they would support the establishment of a marine protected area (MPA) covering most of the New Zealand coast out to the 100-meter depth contour, removing commercial set netting, trawl fishing and recreational set

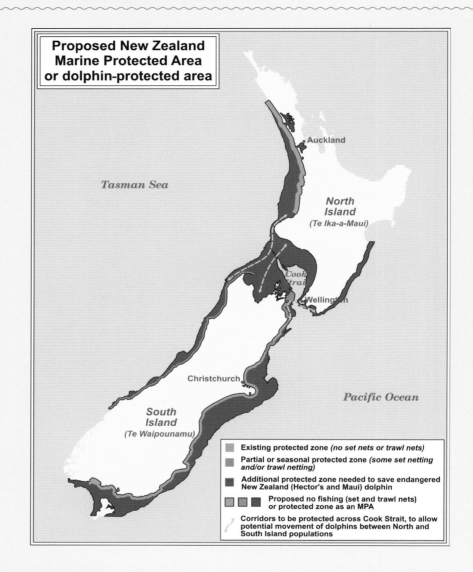

netting from this area. This is the option strongly recommended by the researchers who know these dolphins and who have followed their decline over several decades.

Having New Zealand citizens

behind us is important but not enough. We are working with local communities to make a strong case to the government to create a permanent safe home for these hobbit-like dolphins. Local incomes from fishing are going to be affected, so the government must address the alternatives, including compensating fishermen and exploring dolphin-friendly fishing techniques. The Hector's dolphin is one of the few dolphin species entirely native to the waters of one country. How can New Zealand, a country that has lost the great moas and other birdlife, not do everything possible to save their own dolphin?

New Zealand Hector's dolphins live close to shore around most of the North Island and South Island. The New Zealand government has made some headway in protecting dolphin habitat but the measures need to be greatly extended. The green areas on the map show areas currently protected; the blue areas are partly protected. Areas in red are unprotected dolphin habitat. Dolphin distribution generally extends to the 100-meter depth contour line. Protection of these dolphins throughout their range is critical for population recovery.

But do MPAs work for marine mammals? Case in point: At Banks Peninsula, off the east coast of New Zealand's South Island, a dolphin MPA was set up in 1988, and the dolphins there have been kept safer from nets. This 1,595-square-mile (4,130 sq km) area has worked effectively to start to reverse the dolphins' decline. However, much larger areas will need to be declared as MPAs to save this species.

New Zealand's endemic Hector's dolphin, above, is in trouble.

Hector's Dolphin

Cephalorhynchus hectori

The smallest dolphin in the world—the largest females are only up to 5 feet 4 inches (1.63 m) long—the New Zealand Hector's dolphins are often referred to as the "hobbit dolphins." Unfortunately, this endearing nickname has yet to motivate the New Zealand government to reverse their steady decline. An estimated 7,200 remain from a population of roughly 30,000 dolphins in the 1970s. These are divided into at least four populations, the most endangered of which is the Maui dolphin found off the west coast of New Zealand's North Island. Only an estimated 50 adults and 15 breeding females are left. A few more mortalities of productive females will spell the end of the Maui dolphin. The other three populations will be the next to go.

Small-sized Toothed Whales

The true river dolphins are four species in four genera, and they are members of four different families. They are the **franciscana (*Pontoporia blainvillei*)**, from the family Pontoporiidae; the **Amazon River dolphin**, or **boto (*Inia geoffrensis*)**, from the family Iniidae; the **South Asian river dolphin (*Platanista gangetica*)**, from the family Platanistidae; and the **baiji (*Lipotes vexillifer*)**, from the family Lipotidae, now regarded as extinct. All four species have numerous small teeth, long pointed beaks and poor eyesight. Some of them have odd body shapes.

The franciscana, while considered a true river dolphin, does not live in a river. Instead, it lives in large estuaries and along the coast of southern Brazil, Uruguay and Argentina. This dolphin is rated Vulnerable by the IUCN Red List.

Another group of so-called river dolphins swim close to the shore in the ocean, although they may also occur far upstream in large rivers. These include the tucuxi, the Guiana dolphin and the Irrawaddy dolphin, among the more full-time river residents and other part-time residents and visitors.

Of the true river dolphins, the Amazon River dolphin is the best known. This poster river dolphin has ungainly flippers, a long beak and bulbous head—all seeming to be out of proportion. Normally white to tan in color, when it is active or swimming through warm water, it turns several shades of pink. In terms of numbers, the Amazon River dolphin is

the healthiest of the river dolphin species.

There are two Asian River dolphin subspecies living in separate South Asian river systems—the Ganges and Indus watersheds. (There is genetic evidence under consideration that suggests they are actually separate species.) The Ganges subspecies (*P. g. gangetica*) is found in eastern and central India, Nepal and Bangladesh. The Indus subspecies (*P. g. minor*) is found in the Indus River Basin in Pakistan; a few also occur in India. The eyes of Asian river dolphins are the most degenerate of all dolphin eyes. These dolphins are functionally blind, the eyes reduced almost to pin pricks. More than most other dolphin species, they rely on echolocation. They are rated Endangered on the IUCN Red List.

Amazon River Dolphin, or Boto
Inia geoffrensis

The Amazon River dolphin, popularly called boto, lives its life thousands of miles from the ocean, facing daily challenges and hazards that most other dolphins will never know. Taking

The Amazon River dolphin, facing page, lives in two major river systems, the Amazon and Orinoco, and six South American countries.

Losing the Baiji

In 1987, Japanese researcher, writer and photographer Hal Sato traveled from her home in Tokyo in search of baiji. She began her quest in Shanghai, at the estuary of the Yangtze River. Chinese researchers warned her that her chances of success were slim. Sato met up with French researcher Jean-Pierre Sylvestre, who shared her interest. Cruising up the river, they at last caught sight of spouts, but found that they were finless porpoises, the other Yangtze River residents that live around Southeast Asia. After weeks of fruitless searching, the pair were invited to the captive facility where Sato and Sylvestre got their first glimpse of two baiji—captives that researchers were hoping to breed at a scientific facility. Enamored by the baiji's delicate beauty and its glimmering white body as it slipped out of the water before diving again, Sato returned to Japan, unsuccessful in her quest to see the baiji in its natural habitat.

In February 1989, Sato returned, this time on her own. She met the Chinese researchers from her first trip, but she was told that it would be difficult to get permission to travel farther up the Yangtze River to search for wild baiji. Sato had arrived

One of the last living baiji was the male Qi Qi, kept captive in the late 1980s in the hope that he would breed with a then immature female. He never did. Researcher Hal Sato described him as almost mystically beautiful, with white cheeks, a long, slender beak, smooth skin and tiny, round ebony eyes.

a few months before the deadly Tiananmen Square protest, and many parts of China were by then off limits to foreigners. She waited in her hotel for days and then weeks, hoping a permit would come. Finally, she

was allowed to join a small Chinese expedition.

With multiple boats combing the river, moving ever farther upstream in the Yangtze, at last, Sato had a brief sighting of six wild baiji. On the final day, the team encountered 10 baiji, including a mother and calf. At last, Sato had seen "free swimming baijis in their natural living environment ... they were more beautiful than I could ever imagine."

Before leaving China, Sato was interviewed by local press and thanked for helping with the searching expedition. As the building of the Three Rivers Dam proceeded over the next decade, however, conditions in the river worsened. The captives died without breeding. Wild baiji "sightings" consisted of individuals killed accidentally by fishing gear.

In late 2006, an international research expedition went in search of baiji. They brought an acoustic team equipped with hydrophones to pick up dolphin sounds, as well as some of the world's best dolphin identification eyes. But they were too late. Baiji had been sacrificed to the population pressures and industrialization of China. A few weeks after their expedition, they had to report to the scientific community that the species was "probably extinct." The IUCN Red List, however, still considers them Critically Endangered, and there have been some undocumented reports of possible sightings. But little if any hope remains that researchers and conservationists will be able to do anything except learn from the awful experience of having lost a species.

advantage of the seasonally flooded forest, it swims far from its usual course, rooting among the trunks of partly submerged rainforest trees. It must rely on its acute hearing and echolocation ability to stay alert to crocodiles or jaguars searching for a meal.

The diet of a rainforest dolphin is diverse, with more than 40 fish species recorded in the boto's diet. The boto sometimes breaks up its prey before eating it, using back teeth that feature special flanges for crushing fish. It is the only whale, dolphin or porpoise species that has the equivalent of molar teeth. All other toothed whale species are "homodonts"—meaning they have only one kind of teeth.

The boto often travels solo or in small groups of two (usually mother and calf) or three. Some groups show site fidelity, though they may range over wider areas during the high-water season.

The boto is rated Data Deficient by the IUCN. Researcher Tom Jefferson estimates that there may be 15,000 botos in the Amazon and Orinoco and the various tributaries. He suggests that the overall numbers are healthy, although individual populations in the upper Amazon of Brazil, Peru and Colombia, as well as a separate population in Bolivia, currently considered a subspecies, may deserve Threatened status. Another population that some researchers consider a separate species is the so-called Araguaian river dolphin, found in the inland state of Tocantins in northeastern-central Brazil. Another subspecies lives in the Orinoco River in Venezuela and Colombia. Identifying individual populations (separate breeding units) is fundamental to conservation, and research will help determine whether these populations are separate species.

There are seven porpoise species, all members of the same family. They can typically be distinguished from dolphins by their smaller size—they are less than 8 feet (2.5 m) long. Other notable porpoise characteristics are a small head with little or no beak and a small triangular dorsal fin (excepting the two species of finless porpoises). Even more diagnostic, but difficult to see, the porpoises have spade-shaped teeth, while dolphins have conical teeth.

The porpoises tend to live along the coasts of various regions of the world except for the **spectacled porpoise (*Phocoena dioptrica*)** and the **Dall's porpoise (*Phocoenoides dalli*)**, which also range into pelagic waters. The enigmatic, rarely seen spectacled porpoise lives off southeast South America and in the subantarctic band around the poles. The spectacled porpoise was thought to be all black until researchers saw them alive for the first time and realized that black was just the death color. The Dall's porpoise is the largest porpoise, and it is found across the North Pacific. Traveling in groups of two to 12, these dolphins are hyperactive and extremely fast, throwing up "rooster-tail" splashes wherever they go.

Porpoises live in small social groups and are thought to have a simpler social structure than dolphins, but that may be only that they are elusive and boat-shy and therefore difficult to study and photograph. Some field researchers have used hydrophones in an effort to understand more about their behavior. In many habitats, porpoises are at constant risk of ending up caught and suffocating in gillnets. Researchers are always on the lookout for strandings. As disappointing as it is to find dead porpoises, most researchers will stop everything to examine a carcass for cause of death, clues about biology and diet and to take genetic and contaminant samples.

The best-known porpoise is the **harbor porpoise (*Phocoena phocoena*)**, but the Critically Endangered **vaquita (*Phocoena sinus*)**, or Gulf of California porpoise, native to Mexico, also receives considerable publicity. Mexican and international scientists and conservation groups are working with the Mexican government to try to save this species from extinction.

The **Burmeister's porpoise (*Phocoena spinipinnis*)** resides along the inshore coast of South America south of the equator to the tip of Tierra del Fuego. Its unique, back-swept dorsal fin is covered in tubercles—odd protruding growths along the top edge of the dorsal fin.

The two species of porpoises with no dorsal fins are the **Indo-Pacific finless porpoise (*Neophocaena phocaenoides*)** and the **narrow-ridged finless porpoise (*N. asiaorientalis*)**, both rated Vulnerable on the IUCN Red List. The Indo-Pacific finless porpoise is familiar to boaters and fishers in tropical Asia from the Persian Gulf to China, including the western Indonesian archipelago. This species was the only finless porpoise until it was split in two, in recognition of the narrow-ridged finless porpoise. The range of the narrow-ridged species extends from the Yangtze River to coastal China, South Korea and Japan.

It is difficult to see finless porpoises in the field—especially when they are traveling in small groups—unless they make a fuss. Sometimes hundreds of finless porpoises splash and churn

the water as they speed along inside the Kuroshio, or Japan Current, that runs along the Japan coast. In Japan, these dolphins are called sunameri, and they have a small commemorative marine protected area. This tiny MPA, however, is too small to protect the porpoises; they have dwindled in number in Japan's interior Seto Inland Sea.

In the Yangtze River, the narrow-ridged finless porpoises have been known to do tail stands. Do they do this because the water is becoming too dirty to see anything underwater? Is it an example of a cultural behavior, something learned and passed on only by this population living in the river? On rare occasions, they may swim close to

Resident to the North Pacific, the Dall's porpoise is abundant but has been killed in large numbers as bycatch in high-seas drift-net fisheries. These fisheries have now been banned, yet combined removals by Japanese and Russian fishermen—some intentional, some bycatch—continue to kill about 20,000 porpoises a year.

a boat and can be glimpsed underwater. Agile, they are able to turn their heads from side to side and glance behind them, an ability shared by the beluga and the Irrawaddy dolphin. The Yangtze River has already seen the baiji driven to extinction; researchers are skeptical about the chances of the narrow-ridged finless porpoise surviving in this busy, polluted, heavily dammed river.

Harbor Porpoise

Phocoena phocoena

Relatively small at up to 6 feet (1.8 m) long, the harbor porpoise lies at the other extreme from the big whales. Compared with the blue whale, which reaches lengths of up to 108 feet 2 inches (33 m), and the bowhead, whose lifespan might be 200 years, the harbor porpoise is like a tiny mayfly, racing through its short life, with most living only for a decade.

In 1989, I investigated "New England's Harried Harbor Porpoise" for the magazine *Defenders*. My research turned up an unprecedented number of harbor porpoise deaths in nets. The porpoises were being quietly swept off the decks of New England fishing boats. Increasingly, porpoises and other smaller dolphins and whales suffer bycatch losses, but the awareness is becoming much greater. While people are busy studying and worrying about North Atlantic right whales, humpbacks, blue and fin whales, the little harbor porpoise also deserves attention.

As it turns out, the harbor porpoise is much more harried than we first thought in 1989. In recent years, gray seals in the North Sea along the mainland northern European coasts have been ambushing porpoises, grabbing them by the neck, sometimes drowning them and then eating them. These occurrences are rare, and it seems to be a new phenomenon. Meanwhile, in North Sea waters around the U.K., hundreds of porpoises have been killed by bottlenose dolphins. The dolphins, however, are not eating them. It has been suggested that this intraspecific aggression is related to competition over food resources, but the two species share very few food items. Another theory is that the male bottlenose dolphin may mistake the harbor porpoises for the like-sized bottlenose dolphin calves. They may be eliminating them, just as male lions kill the offspring of other males, so that the females will come into estrus and be available sooner for mating. But both of these ideas are speculative.

Harbor porpoises may get along better with Dall's porpoises, sharing habitat with them across the North Pacific. Harbor porpoise males are known to mate with the slightly longer, heavier Dall's porpoise females. Together, they produce porpoises that hybridize their parents' appearance and behavior.

The two names for this porpoise—"harbor" and "common"—are telling. Its nearshore habitat all along the North Pacific and North Atlantic continental coasts once rendered it a frequent resident as well as common to the harbors. Yet in recent decades, it has been largely absent from busy, polluted harbors, noisy with traffic and other developments. For the short-lived harbor porpoise, life is too brief to spend in a polluted world full of boat traffic. If this species becomes common again in harbors, it will be a sign that pollution levels have indeed improved. As it happens, harbor porpoises have been showing up under the Golden Gate Bridge across San Francisco Bay—an indication that the waters there have become cleaner and that there are fish to catch.

One solution to keeping harbor porpoises around is to protect coastal habitat, where they are often the most vulnerable. In Europe, a number of special areas of conservation have been designated for porpoises, particularly in German, Danish and Irish waters and, more recently, in U.K. waters. The essential next step is to put in place effective management that addresses threats to porpoises in these areas.

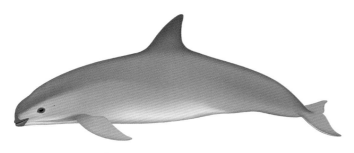

Vaquita

Phocoena sinus

Shy, rare and cryptic are words often used to describe porpoises, but they are never more fitting than when used to describe the Gulf of California porpoise, the vaquita. This porpoise is most often seen dead in a fishing net. For decades, fishers, researchers and the Mexican government have known that this smallest porpoise is in steep decline.

The vaquita was named by science only as recently as 1958. Through the 1980s and 1990s, several surveys searched for it and found numbers were crashing. The Mexican government protected a large part of its habitat, but the net deaths continued. Finally, a dedicated, determined expedition in 2008 used passive acoustics to find the porpoise and to see whether conservation actions to reduce the level and area of gillnetting were enabling vaquita numbers to increase. It also allowed filmmaker Chris Johnson to photograph and film these porpoises in their natural habitat—revealing their lives in greater detail than ever before. But the expedition's conclusion was no surprise: The world's smallest, most endangered porpoise is in serious trouble.

Yet this is a tale of not one but three imminent extinctions. Two are species—the totoaba fish and the vaquita that are caught in the same nets. The other extinction will be the livelihood of the totoaba fishers. Already many have taken government buy-outs aimed to help protect the vaquita. In 2015, with new vaquita population estimates dipping to a mere 97 individuals, the Mexican government declared a two-year ban on gillnet fishing throughout the species' entire habitat in the northern Gulf of California. It also announced enforcement measures. Since then, however, a Greenpeace survey revealed that there were still gillnets in use, confirmed by Sea Shepherd's work to enforce the ban. In July 2016, with the population dropping to 60, Mexico announced a permanent ban. As of 2017, about 30 were left.

The Mexican government has made huge efforts to save the habitat, but has there been sufficient enforcement? Are fewer than 100 individual vaquitas enough to turn the species around?

Nothing is certain except this: Without strong enforcement of the fishing ban, the careful maintenance of a healthy habitat and a great deal of luck, this account may be the last you'll read of the vaquita as a living species. Will this generation witness one of the most notable survival stories of modern species under threat—or yet another extinction?

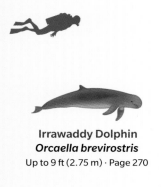

Irrawaddy Dolphin
Orcaella brevirostris
Up to 9 ft (2.75 m) · Page 270

Killer Whale, or Orca
Orcinus orca
Up to 32 ft 2 in (9.8 m) · Page 269

Australian Snubfin Dolphin
Orcaella heinsohni
Up to 8 ft 10 in (2.7 m)
Page 271

Short-finned Pilot Whale
Globicephala macrorhynchus
Up to 23 ft 7 in (7.2 m) · Page 270

Long-finned Pilot Whale
Globicephala melas
Up to 22 ft (6.7 m) · Page 271

False Killer Whale
Pseudorca crassidens
Up to 19 ft 8 in (6 m) · Page 272

Pygmy Killer Whale
Feresa attenuata
Up to 8 ft 6 in (2.6 m)
Page 272

Melon-headed Whale
Peponocephala electra
Up to 9 ft 2½ in (2.78 m) · Page
273

Risso's Dolphin
Grampus griseus
Up to 12 ft 6 in (3.8 m) · Page 273

Guiana Dolphin
Sotalia guianensis
Up to 6 ft 1½ in (1.87 m)
Page 274

Tucuxi
Sotalia fluviatilis
Up to 4 ft 11 in (1.49 m)
Page 274

Common Bottlenose Dolphin
Tursiops truncatus
Up to 12 ft 6 in (3.8 m)
Page 275

Indo-Pacific Bottlenose Dolphin
Tursiops aduncus
Up to 8 ft 10 in (2.7 m)
Page 275

Indo-Pacific Humpback Dolphin
Sousa chinensis
Up to 8 ft 10 in (2.7 m)
Page 276

Indian Ocean Humpback Dolphin
Sousa plumbea
Up to 9 ft 2 in (2.8 m)
Page 276

Australian Humpback Dolphin
Sousa sahulensis
Up to 8 ft 10 in (2.7 m)
Page 277

Atlantic Humpback Dolphin
Sousa teuszii
Up to 9 ft 2 in (2.8 m)
Page 277

Rough-toothed Dolphin
Steno bredanensis
Up to 9 ft 2 in (2.8 m)
Page 278

Pantropical Spotted Dolphin
Stenella attenuata
Up to 8 ft 6 in (2.6 m)
Page 278

Atlantic Spotted Dolphin
Stenella frontalis
Up to 7 ft 7 in (2.3 m)
Page 279

Spinner Dolphin
Stenella longirostris
Up to 7 ft 9 in (2.35 m)
Page 279

Clymene Dolphin
Stenella clymene
Up to 6 ft 6 in (1.97 m)
Page 280

Striped Dolphin
Stenella coeruleoalba
Up to 8 ft 5 in (2.56 m)
Page 280

Short-beaked Common Dolphin
Delphinus delphis
Up to 8 ft 10 in (2.7 m)
Page 281

Long-beaked Common Dolphin
Delphinus capensis
Up to 8 ft 6 in (2.6 m)
Page 281

Fraser's Dolphin
Lagenodelphis hosei
Up to 8 ft 10 in (2.7 m)
Page 282

White-beaked Dolphin
Lagenorhynchus albirostris
Up to 10 ft 2 in (3.1 m)
Page 282

Atlantic White-sided Dolphin
Lagenorhynchus acutus
Up to 9 ft 2 in (2.8 m)
Page 283

Pacific White-sided Dolphin
Lagenorhynchus obliquidens
Up to 8 ft 2 in (2.5 m)
Page 283

Dusky Dolphin
Lagenorhynchus obscurus
Up to 6 ft 11 in (2.1 m)
Page 284

Hourglass Dolphin
Lagenorhynchus cruciger
Up to 6 ft 3 in (1.9 m)
Page 284

Peale's Dolphin
Lagenorhynchus australis
Up to 7 ft 3 in (2.2 m)
Page 284

Northern Right Whale Dolphin
Lissodelphis borealis
Up to 10 ft 2 in (3.1 m)
Page 285

Southern Right Whale Dolphin
Lissodelphis peronii
Up to 9 ft 10 in (3 m)
Page 285

Commerson's Dolphin
Cephalorhynchus commersonii
Up to 5 ft 11 in (1.8 m)
Page 286

Heaviside's Dolphin
Cephalorhynchus heavisidii
Up to 5 ft 7 in (1.7 m)
Page 286

Hector's Dolphin
Cephalorhynchus hectori
Up to 5 ft 4 in (1.63 m)
Page 287

Chilean Dolphin
Cephalorhynchus eutropia
Up to 5 ft 7 in (1.7 m)
Page 287

 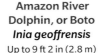  

South Asian River Dolphin
Platanista gangetica
Up to 8 ft 6 in (2.6 m)
Page 288

Amazon River Dolphin, or Boto
Inia geoffrensis
Up to 9 ft 2 in (2.8 m)
Page 288

Baiji
Lipotes vexillifer
Up to 8 ft 6 in (2.6 m)
Page 289

Franciscana
Pontoporia blainvillei
Up to 5 ft 10 in (1.77 m)
Page 289

Dall's Porpoise
Phocoenoides dalli
Up to 7 ft 11 in (2.4 m)
Page 290

Harbor Porpoise
Phocoena phocoena
Up to 6 ft 2½ in (1.89 m)
Page 290

Spectacled Porpoise
Phocoena dioptrica
Up to 7 ft 4 in (2.24 m)
Page 291

Burmeister's Porpoise
Phocoena spinipinnis
Up to 6 ft 7 in (2 m)
Page 291

Vaquita
Phocoena sinus
Up to 4 ft 11 in (1.5 m)
Page 292

Indo-Pacific Finless Porpoise
Neophocaena phocaenoides
Up to 5 ft 7 in (1.7 m)
Page 292

Narrow-ridged Finless Porpoise
Neophocaena asiaorientalis
Up to 7 ft 5 in (2.27 m)
Page 293

7

The Future for Whales

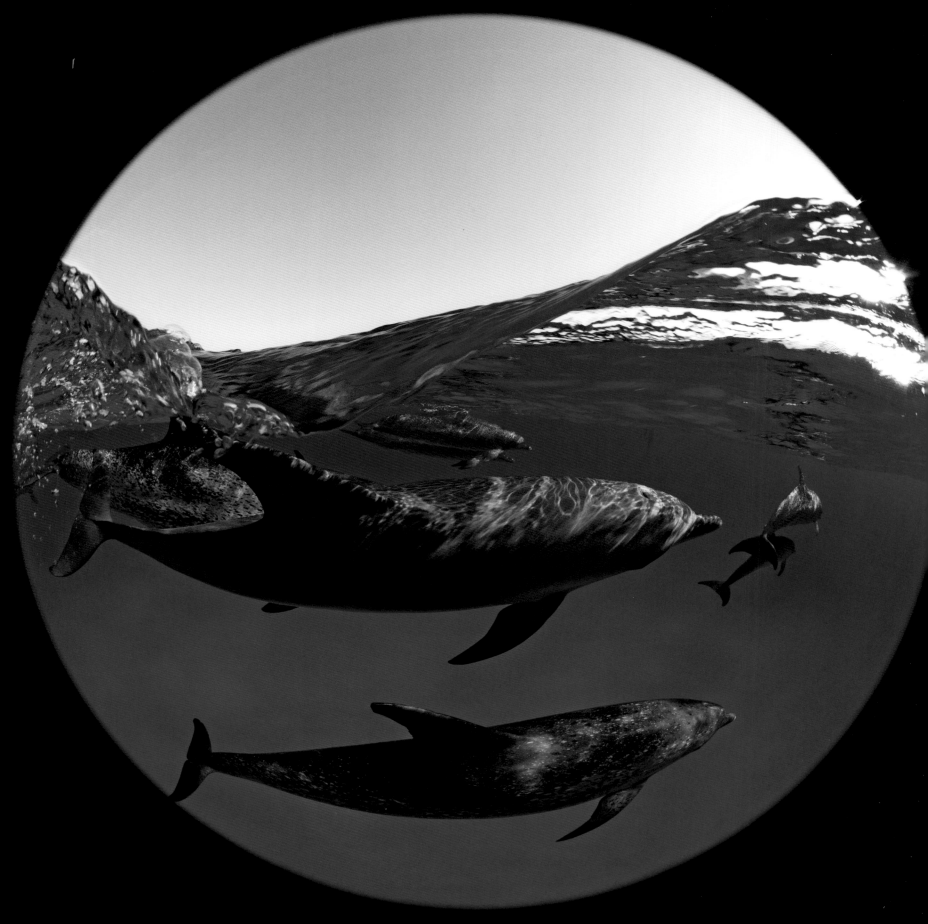

THE FUTURE FOR WHALES
Still at the Edge of Extinction?

I n the 1980s, whale conservationists celebrated as a world moratorium on whaling was put in place. It began in the 1985–1986 Antarctic season—a moratorium that continues to this day. It had all started with the "Save the Whales" movement in the 1970s. With the moratorium established and large-scale commercial whaling at an end, many people thought that whales were finally protected. Yet the legacy from centuries of whaling and the emergence of new threats to many cetaceans has meant that that day is still a long way off.

The latest International Union for Conservation of Nature (IUCN) Red List indicates that the status of more than one third of all baleen whale species remains Endangered (five of 14 species), with four species considered Data Deficient. Nine of the 14 species have at least one subpopulation that is Endangered or Critically Endangered.

Among the 90 species of whales, dolphins and porpoises, the status of 42 (nearly half of all cetaceans) is Data Deficient—we simply do not know enough to be able to say whether populations are healthy or not. The large number of Data Deficient species hints at the difficulties involved in studying whales and reveals how much remains to be learned. Only a quarter of all cetaceans are listed as of Least Concern, meaning species that are well enough studied and can be considered to have healthy populations.

Previous spread: Rough-toothed dolphins are warm-water, usually offshore dolphins and often approach boats and bowride. However, west of the big island of Hawaii, along the Kona coast, where they are known to steal fish off fishing lines and are thus often shot at, they steer clear of boats. Atlantic spotted dolphins, facing page, cruise through the waters of the Bahamas off Bimini and Grand Bahama Island. The younger dolphins tend to have few or no spots; they acquire spots with age.

Critically Endangered, Endangered or Vulnerable—this status alone indicates that conservation measures should be taken. Meanwhile, a precautionary approach may indicate that classifications of Near Threatened or Data Deficient are also grounds for conservation. Thus, in all, about three-quarters of all whale, dolphin and porpoise species may qualify for conservation measures on the basis of their Red List status alone.

In the 30 years since the whaling moratorium began, more than 40,000 whales have been killed in whaling that is being conducted largely outside the "rules." While this represents a fraction of the number taken during the peak whaling years, some of the whales now being killed are threatened species. Japan has killed 18 Endangered fin whales and 1,087 Endangered sei whales, as well as 56 Vulnerable sperm whales. The fin whales, as well as most of some 13,000 minke whales (not Endangered), have been killed in the Southern Ocean Sanctuary, defying the spirit of internationally agreed accords, including the Antarctic Treaty and the International Whaling

In September 2016, a common minke whale is brought into Kushiro port, Hokkaido, Japan. This marked the beginning of another season in the highly criticized, anachronistic whale research program subsidized by the Institute of Cetacean Research in Japan.

For the other quarter, there are various degrees of concern: Two recently designated species have no status yet, but five are rated Near Threatened, five are Vulnerable, seven are Endangered and two are Critically Endangered, one of which, the baiji from the Yangtze River in China, is considered Extinct after a 2006 expedition failed to turn up a single dolphin. As if in hopes that more may turn up, the IUCN hasn't yet officially pressed the Extinct button.

Thus, for the one quarter of cetacean species with an unfavorable or threatened rating—

The gray whale was the first whale to be protected from whaling, the first whale to be awarded a marine protected area and the first whale to be taken off the Endangered Species List. For decades, observers have watched these whales in the lagoons of Mexico and from the shore along the Pacific coast from California to Alaska.

The carcass of a whale rests on the deck of a whaling ship after being harpooned off the west coast of Iceland.

Commission moratorium. Some of these were killed under Japan's formal objections to the moratorium, but most were killed in the name of so-called "scientific whaling," better described as commercial whaling under a name that serves as a smoke screen. Reputable scientists have declared these whaling operations unscientific after examining the research protocols and noting how few peer-reviewed papers have been produced as a result. The overwhelming conclusion has been that this is an expensive, face-saving effort by the Fisheries Agency of the Japanese government to continue to support its whaling industry, defying the moratorium, no matter the cost.

The cost, in fact, is difficult to account for, given the lack of demand for whale meat within and outside Japan. But consider the cost of the "science" itself, money spent that would go much further if used for other kinds of whale research; the cost to the whales in their lives and future conservation prospects; and the cost to the country's image. When Japan hosted the Convention on Biological Diversity conference in Nagoya in 2010, the contradictions in Japan's ocean policies regarding whales and fisheries, as well as the failure to declare functioning marine protected areas in Japanese waters, were clearly evident. Many people working in Japanese conservation would like to see their government advance into the 21st century rather than remaining in the whaling era, subject to the narrow interests of its powerful fisheries agency.

Meanwhile, Iceland and Norway have steered

their own course in defiance of Europe and the world community's whaling moratorium. Since 1986, Iceland has killed nearly 1,000 Endangered fin whales in the North Atlantic, as well as nearly 600 minke whales, some through so-called scientific whaling, some under a much-debated reservation to the moratorium on commercial whaling. Norway, on the other hand, didn't bother with the scientific-whaling charade. It objected to the moratorium from the start and set its own quotas for how many whales it was allowed to kill off the Norwegian coast. These minke whales are not endangered, but, as conservationists argue, they are not Norway's whales to kill. Some, if not most, of these minke whales are believed to range widely through European waters, including the waters off Scotland, Denmark and Ireland.

The greatest damage to threatened species and the strongest argument against whaling, however, can be found in the era before the whaling moratorium and rests with the Soviet Union. In the 1990s, after the fall of the USSR, brave Russian scientists made public secret logbooks, and additional details have been revealed since then. Of the threatened whales, the following tallies cover the period 1948 to 1979, most of which were killed by Soviet whalers in violation of various IWC regulations: 772 North Pacific right whales, 145 bowhead whales, 1,638 blue whales in the North Pacific, and 157,680 sperm whales. In the Antarctic, some 13,035 blue and pygmy blue whales were taken. In total, in the North Pacific (1948–1979), 194,177 whales of 12 species were taken, while 169,615 were reported. In the Antarctic (1946–1986), 338,336 whales were taken but only 185,768 were reported. Some of those reported were misreported—for example, blue whales were misreported as fin whales because of their small size.

It is astonishing that humans have not yet driven a whale species to extinction, although it's true we've come close. Some whales may yet go extinct, such as the North Atlantic right whale and the North Pacific right whale, along with certain populations of blue whales and bowhead whales. Whaling may have put these whales at low numbers, but other factors, such as bycatch, habitat destruction, chemical pollution, ship strikes or climate change, singly or in combination, may yet finish them off.

For the small cetaceans, ranging from river dolphins and porpoises to oceanic dolphins, some of the most serious threats are bycatch from various fisheries. Chemical pollution is also a factor, and researchers are talking about the climate change effects they are now seeing. Of the small cetaceans, one river dolphin has been driven extinct largely by habitat destruction and harmful fishing methods—the baiji, or Yangtze River dolphin, last seen in the early 2000s. Next in line is the vaquita, with roughly 30 individuals left in the upper Gulf of California and maybe as few as only eight breeding females. After the vaquita, the most likely candidate for extinction is probably the New Zealand Hector's dolphin, which has been reduced from a population of 30,000 in the 1970s to 7,200 today. Both the vaquita and the Hector's dolphin have been dying in fishing nets as bycatch. Efforts are being made to save these species by the governments of Mexico and New Zealand, pressed hard by scientists and conservationists, but will their actions be sufficient and in time?

Cetacean Culture

A bottlenose dolphin searches for sponges to hold at the tip of her beak while foraging in the sand for food, thereby protecting her skin from injury.

A young humpback whale new to the feeding grounds watches seven older individuals blowing bubbles and making bubble clouds below a fish school. Seconds later, they all rush up the water column, mouths open.

A female killer whale calf learns the dialect of the pod from her mother, including the unique way in which her pod vocalizes certain calls that are subtly different from a closely related pod in the same clan. Over the next decade, as the calf matures, one of the pod's calls gradually changes. The calf and her podmates learn the new call.

As researchers have spent year after year in the field and come to know the whales, dolphins and porpoises they study, the evidence of cetacean culture is beginning to be revealed.

Culture in cetaceans, as defined by whale scientists Hal Whitehead and Luke Rendell, is behavior or information that is socially learned and shared within a community. It is distinct from the inherited, hard-wired behavior that can be explained through genetics or simple behavioral changes due to ecological factors.

Feeding behavior considered part of cetacean culture includes a diverse array of techniques that are learned and shared, often through the mothers but sometimes from other whales or dolphins in the group. Killer whales that work together to corral and kill baleen whales, as well as to scavenge carcasses of whales killed by whalers, are sharing feeding methods well known from the whaling era. In more recent revelations, killer whales have been photographed creating waves to knock Antarctic Weddell seals from ice floes, while in Patagonia and the Crozet Islands, killer whales beach themselves to catch sea lions and elephant seals. In the Russian Okhotsk Sea, the Strait of Gibraltar, and other corners of the ocean, killer whales have figured out how to strip fish off lines or pull them out of nets.

While killer whales learn their pod's dialects, male humpback whales share their songs across ocean basins. During the winter breeding season, the songs gradually change, with all the males in each region adopting the changes at the same time. These changes have to do with the pitch or frequency of the notes as well as the overall pattern of the phrases and themes, complex changes that are only possible through social learning.

Cetaceans aren't the only animals with cultural lives. Long-lived elephants enjoy a complex matriarchal society that features collaboration, cooperative interaction and shared knowledge. But the most substantial evidence of cultural lives comes from chimpanzees and other primates. Comparisons between human and chimpanzee cultures have been offered up in recent years, and, as Whitehead and Rendell say, different perspectives have led to predictably different conclusions. While human and chimpanzee cultures share a range of features, human culture goes far beyond chimpanzees in its depth and extent. And, of course, there are aspects of human societies, including material-based technologies, agriculture and economic systems, that are completely absent in chimpanzees, cetaceans or any other

non-human cultures. Still, it is easier to compare humans and chimps than to compare either one with whales, dolphins and porpoises.

One substantial difference in the cultural lives of chimpanzees, humans and cetaceans has to do with the presence, or lack thereof, of opposable thumbs and fingers, which allow animals to use and manipulate objects as tools. In order to avoid being injured by spiny fish or rays, bottlenose dolphins in Western Australia hold a piece of sponge at the tip of their beak when prodding for food. These dolphins are engaged in a kind of tool use. Clearly, however, there is much less room for tool development among cetaceans, which

Humpback whales blow underwater streams of bubbles which form into bubble clouds and bubble nets that surround, confuse and herd small fish. The whales then swim up to surface, grabbing a mouthful. Only certain humpbacks use these techniques.

lack fingers and opposable thumbs. But while chimpanzees are closer to humans in that they are able to use tools, the ability to develop tools that accumulate and build on earlier tools is unique in its complexity to humans.

On the other hand, like humans, both killer whales and sperm whales are much more multicultural than chimpanzee societies, and cetaceans and humans are clearly superior to chimpanzees

in terms of social learning related to vocalizations. This last aspect may be key. Of course, the development of spoken and written language by humans to express diverse and complex concepts has taken *Homo sapiens* far beyond our earliest ancestors.

The current conclusion of Whitehead and Rendell, along with other cetacean researchers, is that cetaceans do not have "sophisticated syntactical language," although they do communicate in many diverse ways. Communicating by sound is well suited to the ocean medium where sound travels 4.4 times faster, as well as much farther, than it travels in the air. For the most part, the diverse patterns of sounds in sperm whales, killer whales and others have to do with establishing relationships and bonding between individuals and groups of whales, as well as finding and catching food. Beyond that, researchers are still at the earliest stage of discovering the possible meaning behind what they sense are impressive cultural attributes in these vocal abilities.

Of course, just because the charismatic and well-studied killer whale, sperm whale, bottlenose dolphin and humpback whale possess culture is not to say that all whales, dolphins and porpoises have culture or that they have it to the same extent. The preferred diets and foraging methods of cetaceans vary, from the diverse hunting cultures of the humpback whale to those of the killer whale. Some whales have deep, dirge-like songs (the blue and fin whale) peculiar to life cycles that span ocean basins, while others are much more diverse and constantly changing (the humpback). Still others have dialects (the killer whale). Other

A big bull killer whale beaches himself to grab a baby sea lion at Punta Norte, Península Valdés, Argentina.

Prey-sharing in Killer Whales

Killer whales are picky eaters. The prey list for the species extends to several hundred fish, many marine mammals, including all the large whales, many dolphins, sea lions, seals and sea otters. It also includes squid, penguins and other seabirds. But in many areas of the world where they've been studied, killer whales, although they could catch and eat anything that swims, tend to eat what their mothers tell them to eat.

In the Northeast Pacific waters of British Columbia and Washington State, the northern and southern communities—two separate fish-eating populations of killer whales—look for chinook (king) salmon, the bigger and fatter, the better. Drone videos of the whales show which whales are not getting enough to eat and which are doing well. Killer whales are sometimes seen sharing their prey. Usually the females share with other females, and the males seem to be left out. Biologist Lance Barrett-Lennard witnessed a mature male carrying a Chinook salmon on his head. Usually, the prey-sharing whale holds the fish or other prey in the mouth, tearing off a chunk and sharing it. This male swam for some distance, keeping the prey balanced on his head, before dropping off the salmon next to his mother.

Spying a Weddell seal on an ice floe in Antarctica, these orcas decided to work together to create waves that would wash the seal into the water. Minutes later, they shared their prize.

To avoid getting bruised, a select group of mostly female common bottlenose dolphins from Shark Bay, Western Australia, have taught each other to use basket sponges when hunting nutritious bottom fish. This cultural example of tool use is being passed down from the mothers.

whales seem to make comparatively less use of vocalizations. Even within a single species, when compared with the fish-eating killer whales, the marine-mammal-hunting killer whales are, unsurprisingly, quiet much of the time so as not to disturb the marine mammals they're hunting. Within the comparatively small groups they travel in, these marine-mammal-hunting killer whales may also have much less to talk about.

Part of the buzz around Whitehead and Rendell's investigations into cetacean culture can be summed up in their statement: "We do not know where the limits of their cultures lie." At the same time, they are stepping back to examine the "evolutionary forces that favor culture as a solution to life's challenges."

Whitehead and Rendell are also looking forward, along with Philippa Brakes and Mark Simmonds in their book, *Whales and Dolphins: Cognition, Culture, Conservation and Human Perceptions* (2011), asking how our understanding and appreciation of cetacean cultures might change the way we think about them. Do cetaceans deserve recognition of their rights?

Getting Close to Whales, Dolphins and Porpoises

The first response that visually oriented humans have when it comes to experiencing animals in the wild is to try to get closer to them. Typically, wild animals prefer to keep their distance, but many cetaceans tolerate these close approaches.

Even more surprising is the extent to which cetaceans approach humans. Early whale watchers were puzzled when gray and humpback whales came close. Didn't the years of whaling count as bad experiences with humans? What about the walls of entangling fishing nets and other gear they must have encountered? Of course, the whales that are killed are removed from the population, but others carry harpoons, shrapnel, as well as pieces of fishing gear with them, sometimes for life. They may also carry the memory of being chased by hunters. Maybe these "experienced" whales avoid humans, while certain others, mostly younger animals, are the "friendly" whales.

A special case of "friendly" behavior, as whale researchers call it, developed in the 1980s with gray whales in the breeding lagoons of Baja California, Mexico. A number of young whales began approaching boats full of whale watchers. Parking alongside the small boats, the whales allowed whale watchers to touch them. Sometimes, they would lift the boat up, providing even more of a thrill for the whale watchers.

A variant of "friendly" behavior can be seen in the seemingly fearless killer whales and bottlenose dolphins. Both of these species have had negative human experiences, ranging from fishermen shooting at them to getting captured for aquaria, with families sometimes split up. On the other hand, killer whales have no natural predators, while dolphins have shark predators to watch out for. Yet, if anything, bottlenose dolphins are even bolder than killer whales in approaching humans.

For many cetaceans, coming close may not be so much friendliness but the curiosity that is the mark of a good predator. To be successful, predators need to inspect new additions to their environment. Yet what defines bottlenose dolphin and killer whale experiences with humans is often bold curiosity. It is beyond what most other dolphins and whales exhibit, although the individual friendly grays, humpbacks and occasional others may be special cases.

What bottlenose dolphins and killer whales have in common is that they both have been extraordinarily successful at adapting to a wide variety of habitats around the world. They have spread out, living in coastal as well as offshore waters. They have adapted so well that separate breeding units, or populations, have formed, leading to new ecotypes and even new species. There are now two accepted species of bottlenose dolphins with another that is probable. While there is still only one accepted killer whale species, scientists agree that killer whale taxonomy needs revision. There may well be several species of killer whales, and some argue that many of the 10 main orca ecotypes known today might eventually be named as species.

Bold curiosity and adaptiveness in bottlenose dolphins may have led to long-term relationships with humans. At the Monkey Mia resort beach in Shark Bay, in Western Australia, certain individual dolphins regularly enter the shallows to mingle

In a longtime mutual association, bottlenose dolphins corral and drive fish toward the fishers in the waters of Laguna, southeastern Brazil. The dolphins get their reward for collaborating by catching escaping fish as the net is closed.

with and be fed by visiting tourists. Hundreds of thousands of people over five decades have now witnessed this phenomenon, with many participating in the feeding. Researcher Janet Mann and others studying this behavior revealed that it was the same six to 10 dolphins and their progeny that were being fed by the public. Mann compared the calves of those mothers being fed with the rest of the population and found they were twice as likely to die in the first year. There may have been more

exposure to sewage or other contaminants in the water or from the provisioned fish. Or were the mothers less attentive to their babies because they were busy interacting with humans? In any case, with stricter controls on feeding and the fish that

Whale watchers gather near Point Vicente Lighthouse, in Rancho Palos Verdes, Los Angeles, to count gray whales on migration. The work of citizen scientists has created a unique record of these whales as they returned from near extinction.

are given, the situation has improved, but there is still some disparity between free-living dolphins and those accepting food handouts. Clearly, feeding wild animals is not a good idea.

Yet there is also something else at play here. Predator curiosity does not completely explain cetaceans coming close to humans. With many whales, dolphins and porpoises, there is an intense need for social contact. This is one of the defining characteristics of social mammals. We see it in solitary dolphins that approach humans in different parts of the world. We see it in young gray whale friendlies and in the bold curiosity of lone bottlenose dolphins and killer whales. We know that this social desire is such that individual dolphins of one species will join up to travel and hunt with a group of dolphins from another species. These may sometimes become long-term relationships.

Could humans join up with a foraging dolphin pod? Could dolphins join up with humans? In cases in Brazil, Australia, India, Mauritania and Myanmar, shore-based human fishermen and bottlenose and other dolphin species have developed a long-term relationship hunting together. Along the mud shores of Laguna in southeast Brazil, traditional fishers wade into the water and wait for the "good dolphins"—the ones they know by name who will help them. The fishers announce their presence by smacking their nets and the palms of their hands on the surface. The bottlenose dolphins proceed to drive the mullet into shallow water. When the fishers see the dolphins arch their backs in a certain way, they know it is time to slip the nets in.

This cooperative fishing in southeast Brazil has been happening as long as anyone is able to remember, a strategy that has been passed down through human fathers and dolphin mothers. Only 30, roughly half, of the dolphin population participates. The probable benefit for the dolphins is that they can grab some of the escaping fish, which are perhaps easier to catch when disoriented or injured by the net. With some variations, this is the way cooperative fishing takes place in other areas of the world too. These dolphin-human fishing cooperatives are an example of extended closeness. Hunger, the hunt for food,

explains a lot. Many species, including humans, will cooperate with unlikely bedfellows in pursuit of a good meal.

The associations between fishers and dolphins have been cherished wherever they occur. But there are also concerns about too much closeness, especially in the expanding whale- and dolphin-watching industry. Commercial whale watching began with gray whales in 1955 in southern California and expanded slowly along the west coast of North America before jumping to Cape Cod on the U.S. east coast and expanding from there to more than 100 countries around the world. When whale watching was young and a few thousand people per year were meeting whales, close encounters were not an issue. Today, however, with more than 13 million people going whale watching and many of them seeking close encounters, there are concerns.

Research in New Zealand, Australia, Canada and other countries has shown short-term impacts on various populations of cetaceans—for example, they may move away from the boat, bunch up and spout more often. Some long-term impacts have also come to light, particularly with small dolphin populations wherein the same dolphins repeatedly have contact with and are disturbed by multiple boats. At its best, whale watching helps promote conservation and local coastal communities, but we must remember that cetaceans need space and time to themselves. We must be careful not to abuse the curiosity, fearlessness and social natures of cetaceans. For the interests of local communities, the businesses that depend on whale watching, as well as for the cetaceans themselves, whale watching must be examined carefully and put on a truly sustainable basis.

Are Whales Dangerous?

In recent years, a number of humpback whales have collided with whale-watching boats, evoking for some the whaling-era accounts of whalers being sent to hasty funerals at sea by such collisions. In fact, there are some similarities between the two scenarios. But evidence suggests that these episodes should be considered accidents rather than examples of the humpback's aggression toward people or boats.

But what about the sperm whale—the largest toothed whale? Big-brained, clever and highly social, the sperm whale was the object of relentless Yankee whaling for most of two centuries. Herman Melville's novel *Moby-Dick; or, The Whale* told the story of an aggrieved male sperm whale bent on destruction of ships and humans. Melville was looking for a good story, and he found it in the *Essex*. In so doing, he largely ignored Thomas Beale's writing in 1839 that sperm whales were "a most timid and inoffensive animal…readily endeavouring to escape from the slightest thing which bears an unusual appearance."

In many thousands of hours spent observing sperm whales as they crossed oceans in small sailboats and cruisers, Hal Whitehead and Linda Weilgart never witnessed an aggressive act from sperm whales. On the contrary, they concluded that sperm whales are indeed shy and frightened more often and more visibly than other whales, retreating into defensive postures when uneasy.

Sperm and other whales, as well as other cetaceans, are said to pass on their culture through imitation of behavior. Strictly hypothetically, Whitehead allows that a small group of sperm whales, having witnessed a male or senior female accidentally overturning a whaling boat and successfully avoiding being killed, might start to imitate this violent behavior. But that's not to say that it has happened or that such an event is likely.

Instead, historical sperm whale "attacks" appear largely to have been caused by the impact of flailing harpooned whales, which led either to boats being overturned or the harpoonist or boat being pulled under water. More than 600,000 sperm whales were killed from the 18th to the 20th century, reducing the species to one third of their original numbers. With thousands of whaling ships approaching close to flailing whales, it would not be surprising that some whalers died during such a dangerous enterprise. Similarly, with thousands of whale-watching boats in operation worldwide and more than 13 million people going whale watching every year, accidents have happened. A few of them have been accidents fatal to humans.

What about the killer whale? Killer whales have injured a number of trainers and killed several in aquaria. This is thought to be due to the abnormal, socially deprived or disturbed conditions in captivity that might lead a bored killer whale to "play" too rough with a trainer. It's significant, however, that no killer whale has ever harmed a human in the wild, despite many in-water and close encounters with researchers and filmmakers over half a century.

Perhaps most surprising is that so few cases of documented, supposedly aggressive attacks exist. Compared with elephants or other animals, cetaceans are extraordinarily passive. An attempt by humans to separate a female from a young calf may result in exceptional behavior, but even so, the evidence for Moby-Dick-style revenge is non-existent.

Whales, such as these humpbacks, rarely present a danger to ships, but ships sometimes hit and injure or kill whales. Whale-watching skippers are careful, though accidents happen.

The Whale Trail

Land-based whale watching has a rich history of raising awareness and inspiring the public to become "ocean citizens." As we've said, it started with gray whales in California in the 1940s and spread up and down the Pacific coast to Canada and Mexico along the coastal gray whale migration route. Land-based whale watching then developed in South Africa for southern right whales, in Quebec for blue and humpback whales and in Australia for Southern Hemisphere humpbacks that moved up along that continent's east and west coasts.

In the 1970s, photo-ID pioneer Michael A. Bigg initiated killer whale censuses in British Columbia that resulted in the involvement of members of the public. These whale lovers were mainly watching killer whales from coastal lookouts, where they were identifying and reporting killer whale sightings and then sending in photos to be identified.

In 2003, the southern community of fish-eating killer whales was declared Endangered in Canadian waters. The United States followed suit in 2005. In part, this designation recognized that the killer whale had never recovered from the intensive aquarium captures and that, given the decline in their salmon prey, increasing toxin accumulations and vessel impacts, population trends were not encouraging. Extinction could occur in as few as 100 years.

In 2008, software professional Donna Sandstrom, who had helped facilitate the return of the orphaned killer whale Springer to her pod off northern Vancouver Island a few years earlier, started a non-profit organization called The Whale Trail to build awareness of and inspire stewardship for the southern resident orcas. Working with a team of partners that includes NOAA Fisheries, the Washington Department of Fish and Wildlife and the Whale Museum, the original goal was to identify sites around the region where it was possible to view orcas or other marine mammals from shore. All sites have a page on The Whale Trail website (thewhaletrail.org), and many feature interpretive panels that have been customized to show which animals can be seen and when. Distinctive Whale Trail markers identify other whale-watching locations.

From 16 initial sites around Washington State, The Whale Trail has expanded up and down the west

coast to British Columbia, Oregon and California—throughout the southern resident orca range and beyond.

The group also organizes frequent talks and community events—all with the goal of connecting coastal residents to the southern orca community and other marine wildlife and their habitat. They're also trying to inspire people to take action to help preserve the marine environment and thus ensure a future for both whales and people.

Essential whale-watching tips

From land: Bring binoculars, a long lens, a tripod, a marine-mammal identification guide, a thermos flask, water, snacks and patience. Be careful not to disturb the site; leave it just as you found it.

On a commercial whale-watching boat: Find a trip that has been recommended and has a good naturalist guide. During or after the trip, do express any concerns you have about the trip to the naturalist in the first instance, and if necessary, follow up by approaching the operator, company and local authorities. Do your research, and try to select an area that is considered low impact/low risk. Medium- to high-impact/high-risk areas are characterized by three or more boats watching the same small coastal population or populations of cetaceans in a confined geographical area, especially if those cetaceans are static (resting, feeding or socializing). Thus, low-impact/low-risk areas would have only one or two boats on whale-watching expeditions in the open ocean.

On ferries and cruise ships: Keep your eyes peeled. Have a word with staff, and ask whether cetacean sightings are announced to passengers and about the best places from which to watch during the journey.

On a private boat: If you see whales, the best thing is to slow down, put your boat into neutral if it is safe to do so, and watch from a distance. Do not make head-on approaches or cut off whales by pulling in front of them. Commercial whale-watching boats in most countries follow guidelines, codes of conduct and regulations, and private boats that encroach upon the whales can be prosecuted. If the whales come to you, stay calm, and keep the boat in neutral. If you are traveling and dolphins come to bowride alongside your boat, maintain your speed and steady direction. When slowing down, do so gradually.

Do not try to touch or feed cetaceans. Do not jump into the water with them. Remember that this is their habitat, their home. Treat cetaceans and their homes with respect.

One of the best spots for watching killer whales in the United States is Lime Kiln State Park, on San Juan Island, Washington, part of The Whale Trail. The coastal waters have been proposed as a whale protection zone, a small area where boats would be excluded.

Pollution: How We Can All Help

What happens when we flush the toilet? Or throw our garbage in or near the water? Or allow industries to use a river, a bay or the open ocean to dispose of their chemical wastes? What happens to the big-city waste from thousands of restaurant deep fryers emptied into storm sewers that then overflow with an oily poison cocktail?

Since humans evolved to live on Earth, we have asked the ocean to absorb our waste products and wash them away. Chemicals from farms and industries are poured or leak into rivers and flow into the sea. In the past, nuclear power plants, chemical companies and governments have used the ocean to dump their waste. Things are improving. Yet the waste from many coastal cities is still shipped by barge to be dumped at a point farther and farther from shore. Plastics of all kinds—including plastic bags—are floating around in the oceans, where they may never degrade or decompose.

And yet there is only one ocean, and all the water circulating today is the same water that existed thousands of years ago and that will exist thousands of years from now. It is a closed, finite system. Water may be the Earth's most precious resource. We pollute the ocean at our peril.

Plastic bags and other ingested substances have contributed to the death of whales due to intestinal blockage leading to starvation. While the

Marine debris researchers estimate that there are 5.25 trillion pieces of plastic debris in the ocean. Even the coastlines of remote Pacific islands can be thick with debris that kills marine mammals, birds, turtles and may be ingested by the fish that end up on our tables.

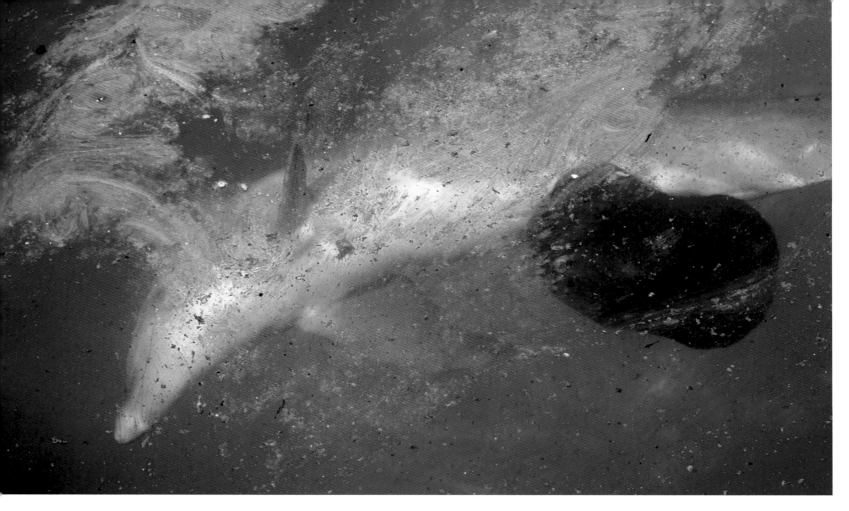

Pollution can be invisible or it can be clearly seen, as in this photo from Curaçao of a common bottlenose dolphin swimming in an oil slick, a black plastic bag trailing across its belly.

precise cause of death may be uncertain, stomach studies still tell a sad tale. A male Cuvier's beaked whale washed up on the Isle of Mull, Scotland, with 23 plastic bags in his stomach. In 2010, a gray whale stranded alive on a beach in West Seattle. After the whale died, the stomach was opened to reveal more than 20 plastic bags, plastic pieces, towels, surgical gloves, a pair of sweatpants, duct tape and a golf ball.

Most ocean pollution, however, is invisible. The legacy of PCBs, used in refrigerants and industrial applications, as well as dioxins and other chemicals, has produced a lethal cocktail for whales living along coastlines and in inland waterways. Many of these chemicals persist in the environment even though they have been banned for decades. A 2014 study voiced concern about the volume of plastics in the world ocean—it was estimated to be 5 trillion pieces of plastic that weighed nearly 297,000 tons. Most of it had broken down into pieces less than 1 mm in length, becoming so-called microplastics that never dissolve or disappear from the ocean. Instead, the tiny pieces were found to be settling in among corals and deep-sea sediments at volumes 1,000 times higher than those near the surface of the sea. These microplastics are getting into the food chain, into fish and crustaceans, and then they are moving up the food chain. But the extent of the impact on marine life and human health is unknown.

The fish-eating killer whales, the southern orca community, are based around southern Vancouver Island and in Puget Sound but range down the west coast of the United States as far south as central California. In 1974, after seven years of intensive captures, they numbered 70 individuals. Before that, there are no precise numbers, but some say that there may once have been as many as 200 individuals in this population. As of 2016, there were 85 individuals, including nine new calves born in the previous year, but in recent years, there have been quite a few deaths of calves. Analysis of the blubber and other tissues from some of those that died has found that they were carrying PCBs and other dangerous chemicals. Mature females may look healthy even though they carry these chemicals in their blubber. However, when their chinook salmon food is in short supply, they start metabolizing the body fats that store all the toxins. If the females are pregnant, this may lead to the loss of the fetus, and if they have recently given birth, then toxin-laden milk may be fed to the calf. In 2014, the southern community lost four orcas, including a 19-year-old productive female called Rhapsody, whose near-term fetus died first but could not be expelled.

Were some of these deaths due to pollution or knock-on impacts from a shortage of their preferred salmon prey? We're still learning, but we do know that southern resident orcas are among the most contaminated marine mammals in the world.

Whale savers come in all sizes, shapes and professions. They can be families walking along the beach doing a weekend beach cleanup, picking up plastic bags and bottles and other garbage so it doesn't go into the ocean. They can be coastal residents or visitors who report an onshore stranded or entangled cetacean.

Whale savers can be fishermen who use careful line-based fishing methods instead of the killing set nets or the trawls that scrape the bottom. Set nets and bottom trawls are destructive to all life, catching many other species besides the target fish, including whales, dolphins and porpoises.

Whale savers can also be mariners on pleasure boats or racing yachts who post a watch and slow down to 10 knots or less in known or suspected whale areas so that they don't strike and kill a whale. Particularly susceptible to ship strikes are right, humpback, fin, blue and sperm whales.

Whale savers can be farmers who reduce or eliminate the herbicides and pesticides that they introduce and are careful with upstream water contamination.

Saving a whale, dolphin or porpoise can start out as an act of compassion for a fellow creature, but it is also an essential act for human survival. Cetaceans ensure the functioning of ocean ecosystems. Tethered to the surface by their need to breathe, cetaceans are the most visible indicators we have in the ocean. If cetaceans look healthy, are reproducing and have populations that are stabilized or increasing, then the ocean is likely to be healthy too. If they are stranding on beaches in increasing numbers, as they did in the years after the Gulf of Mexico Deepwater Horizon BP oil spill, then we know things are not right. Whales, dolphins and porpoises are an early warning system—the best clue we have to the health of the seas.

Noise as Pollution

A kind of shouting war is happening in the ocean these days. Humpback whales and killer whales, among others, command the sound spectrum that largely matches human hearing. Dolphins work the higher ranges, from human-audible up to so-called ultrasonic. Blue whales and fin whales live mostly in the infrasonic, below human hearing. At the same time, noise from ship traffic, navy sonar, hydrocarbon and seismic testing also cover most of this broad sound spectrum. The louder the human noise becomes, the louder the cetaceans have to "shout" in order to communicate with each other.

Noise in the sea can be classified as chronic and acute. The chronic aspect refers to the background or ambient noise levels that at the lower frequencies have increased by an average of 3dB to 4dB (decibels) per decade since the 1950s, based mainly on studies in the North Pacific. Decibels measure sound intensity on an exponential scale. Thus, the sound intensity of this background noise has roughly doubled every decade over a period of time roughly comparable to the lifetime of a whale. The increase in chronic background noise is thought to come largely from the growing number of container ships, some 50,000 of which now ply the world ocean. The impact of this increasing background noise is difficult to measure, but some scientists are concerned that the noise may be masking cetacean communication, which in effect reduces habitat. It may also limit their ability to hear ships or lead to habituation, which in turn may cause an increase in the number of collisions between ships and whales.

Whales, dolphins and porpoises produce a wide variety of sounds that vary by species as well as in their function in finding food, navigating and communicating with each other. The ability to hear in a watery habitat is critical. Vision is limited under water, reduced to 30 feet (9 m) or less in plankton-rich temperate and polar waters to a little more than 100 feet (30 m) in some tropical areas. Sound, on the other hand, travels at speeds that are 4.4 times faster in water than in air. Depending on frequency, these sounds can permeate many square miles; in the case of low-pitched sounds, they can travel for hundreds or even thousands of miles. Singing male humpback whales fill many tropical waters in winter with their long and complex songs. Killer whales, on the other hand, have loud, piercing calls when they hunt salmon or other fish but remain quiet when hunting other marine mammals or whales. Blue whales, the greatest ever long-distance communicators, send their bellowing sounds across ocean basins.

But what happens when the ocean itself becomes noisy? Noise, like whale sounds, also carries over great distances. Before modern shipping traffic, loud noises at sea were confined to underwater volcanic explosions, tsunamis, meteors hitting the water—infrequent natural phenomena. Of course, there was background noise from rain and hailstones hitting the surface and the action of waves crashing, as well as the ever-present snapping shrimp and other marine life and the sounds of marine mammals themselves.

But alongside, adding to chronic noise, there is acute noise. Primary causes of it are navy sonar, explosions from hydrocarbon exploration and seismic testing using air-gun arrays.

Navy sonar from the United States and NATO are thought to have been responsible for a number of mass strandings of beaked whales, especially Cuvier's beaked whales. In the waters off Greece in May 1996, 12 Cuvier's beaked whales stranded alive and died at various locations along the coast. At least four of them were found to be bleeding from the eyes. The event occurred right after NATO exercises involving the use of loud mid-frequency sonar.

Four years later, in March 2000, in the Bahamas, several U.S. Navy warships were using sonar over a protracted period. In the course of two days, 14 beaked whales, including nine Cuvier's beaked whales, were found stranded along the Grand Bahama and Abaco shoreline, most of them still alive. Some were pushed back to sea; others died on the beach. Researchers Ken Balcomb and Diane Claridge witnessed these strandings. Pathologists later uncovered "some type of auditory structural damage ... specifically bloody effusions or hemorrhage near and around the ears." They concluded that while these injuries would not have been fatal, they might have been part of a sequence that eventually led to the whales' deaths. Initially, U.S. Navy officials accepted no blame, but after a joint investigation by the Navy and NOAA, they concluded that the use of naval mid-frequency active sonar (MFAS) had indeed precipitated this beaked whale mass stranding.

In 2004, in the Canary Islands, another NATO exercise coincided with a mass stranding of

The Cuvier's beaked whale is a deep-diving whale susceptible to navy sonar. This adult male stranded at Ostend, Norfolk, on the British North Sea coast. Mass strandings have occurred in the Bahamas, Greece and the Canary Islands.

Cuvier's beaked whales. This led researchers who had been looking into historical strandings among Cuvier's and other beaked whales around the Canary Islands to consider whether they, too, had occurred as a result of naval events. After consultation, a voluntary 50-nautical-mile (92.6 km) exclusion zone was created around the Canary Islands for naval exercises, and since then, no further strandings have occurred.

More recently, Mediterranean governments asked the region's scientists to investigate the issue of noise and cetaceans. The researchers proposed implementing an exclusion zone for navy sonar and seismic surveys similar to

The cavitation from poorly performing ship propellers is a main cause of chronic noise in the sea.

that in the Canary Islands, to protect the last remaining Cuvier's beaked whale habitat in the Mediterranean. The governments of France, Greece and Cyprus, however, have blocked the proposal.

A number of scientists have explored why whales might be stranding and dying from loud sonar and other noise, as well as why the victims are mostly beaked whales, particularly Cuvier's beaked whales. The most plausible explanation comes from two cetacean veterinary specialists, Professor Antonio Fernández from the University of Las Palmas, Canary Islands, and Paul Jepson from the Zoological Society of London. Fernández and Jepson have identified "acute and systemic gas and fat embolism" in mass stranded beaked whales—a condition associated with a suite of tissue injuries similar to decompression sickness (DCS), or "the bends" in human divers. These specialists have concluded that this syndrome in deep-diving whales might be caused when they surface too swiftly, perhaps in response to being startled or frightened by acute noise. Compared with other whales and dolphins, a beaked whale startled while feeding deep might be the most susceptible to the bends.

Researchers and conservationists are at an early stage of understanding the impact that noise in the ocean is having on whales, as well as on other species such as fish and invertebrates. Many scientists argue that it is time for the world's navies, hydrocarbon and other industries and the International Maritime Organization (IMO), which regulates shipping traffic on the world ocean, to adopt a precautionary approach by trying to reduce human-generated noise. The Convention on Biological Diversity (CBD), to which most of the world's countries have agreed, has an abiding interest in the proliferation of noise on the high seas, while the Convention on Migratory Species (CMS) is concerned about the noise impacts on whales because of their dependence on sound during many of their activities, including migration. Both the CBD and CMS are looking at spatial solutions. Researcher Rob Williams recommends the adoption and maintenance of quiet, or quieter, marine protected areas (MPAs). More efficient propellers and slower speeds in areas of high whale abundance would help. Along with the CBD, CMS and IMO, the coastal nations of the world need to take action in their own waters as well as regionally and on the high seas so that noise does not degrade and destroy biodiversity in the ocean.

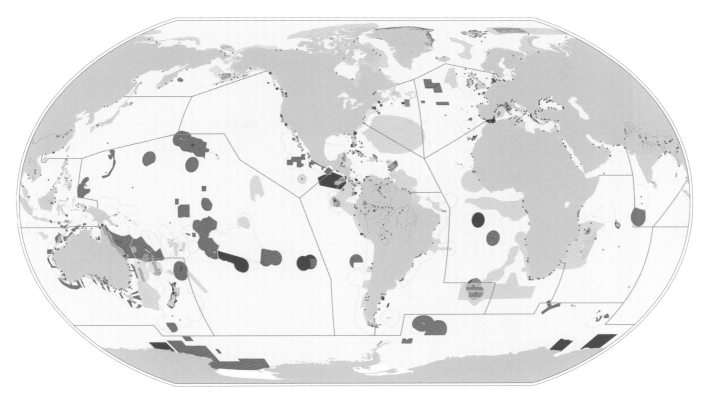

⬤ · Existing Marine Protected Areas (MPAs) with cetacean and other marine-mammal habitat.

⬤ · Proposed MPAs with cetacean and other marine-mammal habitat.

Ecologically or Biologically Significant Areas (EBSAs) identifying marine-mammal habitat through the Convention on Biological Diversity.

Biologically Important Areas (BIAs) with marine-mammal habitat (in Australia and the United States only).

Grey lines surround countries that have proclaimed their waters as "no hunting areas" for cetaceans.

Lighter lines indicate the boundaries of the 18 IUCN marine regions.

Homes for Cetaceans: Protecting Their Habitat

To ensure healthy populations of whales, dolphins and porpoises, it is essential to protect their habitat. A marine protected area, or MPA, is a catchall term that describes sanctuaries, marine national parks, special areas of conservation, and so forth, each of which has designated rules and levels of effectiveness and enforcement depending on the country and stakeholder involvement. Good MPAs are able to accommodate regulated

human uses of the ocean at a modest level, while focusing on long-term nature protection. It is a challenge, and to meet it, MPA managers use a technique called adaptive management, learning and adjusting as they go along.

An important and visionary type of MPA is a marine reserve offering a high level of protection without commercial fishing and other development activities. A marine reserve is sometimes

The Story of Billie, or How to Set Up a Marine Protected Area or Sanctuary for Dolphins

Dolphin researcher Mike Bossley, as a scientist and conservationist with Whale and Dolphin Conservation and formerly Greenpeace, had watched the evolution of whale conservation from "Save the Whales" to "Whales won't be saved until we save their habitat." Eventually, Bossley mobilized a community in Australia to set up a dolphin sanctuary in the industrial slum of a river in Adelaide, the busiest port in South Australia. He photo-identified his dolphins and came to know their stories. His poster dolphin, a young female, turned out to be the unlikely "Billie."

Orphaned at about two years of age, Billie lived in a section of Adelaide's highly polluted Port River estuary. She kept company with another dolphin, an old male mistakenly dubbed Big Mama. In late 1987, Billie left the river and swam some 12 miles (20 km) south along the coast, where she became entrapped in a boat harbor. Rescued and transported to a local aquarium that kept dolphins, she recovered her strength, and as she did, she observed the performing dolphins in the show. After several weeks, she was released and returned

A dolphin named Wave shows off his tail walking in the Adelaide Dolphin Sanctuary.

to the river. She thrived as she matured and began to have calves.

Roughly seven years later, Bossley documented Billie doing something unusual for a wild dolphin—"tail walking." Tail walking is an aquarium stunt performed by trained captive dolphins. The dolphin rises up high out of the water and moves backward as if walking on its flukes. Three years later, Billie was seen tail walking again, but there were no further documented tail walks until a decade later, in 2007. This time,

the stunt was performed by Wave, a mature female unrelated to Billie. The news that Billie's wild mates were imitating her tail walk spread around Australia and the world. Wave became a champion tail walker, as well as a tail-walking teacher. At least 10 other dolphins learned how to perform the tail walk by watching Wave—all wild females and calves. Tail walking appeared to be of little interest to adult males.

As time passed, Bossley realized that although Billie was reproducing, all but four of her six calves had died. One of the surviving calves narrowly escaped death when he was sucked into a factory cooling sump. As Bossley shared stories about Billie with the public, he described the polluted environment in which Billie and the other dolphins lived and argued that it was critical to create a cleaner, safer home for the dolphins.

In all, Bossley had counted at least 30 Indo-Pacific bottlenose dolphins living in the river, but some 300 transient dolphins would occasionally visit. The river was a polluted industrial zone, yet it still had some mangroves, seagrass, saltmarsh, tidal flats and creeks that provided food resources. In other words, there was still hope for the habitat.

Over several years, the local community, driven by Bossley and his dedicated team of researchers and citizen scientists, appealed for the cleanup of the river and the creation of a protected area for the dolphins. In June 2005, the Adelaide Dolphin Sanctuary was established with protection for 46 square miles (118 sq km) of the Inner Port to the Outer Harbour and North Haven Marina, stretching north to the mouth of the Gawler River.

Billie was last seen performing her tail-walking routine in 2009. Later in 2009, she became emaciated and was eventually captured for veterinary assessment. She was found to have advanced kidney failure and was euthanized. For a few years afterward, Billie's spirit could be glimpsed every time tail walking was noted. The years 2009, 2010 and 2011 were the three top record years during which researchers noted tail walking among Billie's mates. Since 2012, tail walking is seen less and less.

Billie's real legacy, however, was the designation of the dolphin MPA. Billie had helped Bossley achieve the goal of protection for her mates and for their habitat in the river. Adelaide's industrial estuary and river became the Adelaide Dolphin Sanctuary.

called a "no take" area. Such protection is good stewardship of the ocean and keeps ecosystems intact without human interference. But it has also led to increased fish stocks both within and outside a reserve.

The first whale MPA was declared by Mexico in 1972 for gray whales. The location was the winter breeding grounds of Laguna Ojo de Liebre, also known as Scammon's Lagoon, in Baja California. In the mid-1800s, the whaler Charles M. Scammon had discovered the lagoon, and he and other whalers then caught and killed as many gray whales as they could, though a few whales escaped the whalers' harpoons. Protection of this lagoon and neighboring San Ignacio Lagoon helped launch whale conservation efforts in Mexico. American ships from San Diego started bringing tourists on multi-day tours. Mexicans who lived in the area did not benefit from tourism until these ships were banned from the lagoons. This created an opportunity for locals themselves to take tourists to see the whales using their own small boats, called pangas. In 1988, the three main lagoons and the surrounding land and sea areas were turned into El Vizcaíno Biosphere Reserve.

The developing tourism industry in the lagoons linked local residents to the whales and gave residents a stake in the whales' future. In the 1990s, when Mitsubishi announced that it would be developing one of the lagoons as a big salt works, local Mexicans protested and were soon supported by the wider Mexican society and by international conservation groups. In 2000, after escalating protests, Mitsubishi withdrew its application. People who thought the lagoons were "paper MPAs" now realized that the support of the community—the stakeholders—had made the MPA real.

Gray whales have continued to breed and raise their calves in Baja, migrating to Alaska in the summer months to feed. No longer considered Endangered, their numbers have hovered around 22,000 since 2000. In 2007, Stanford University's Hopkins Marine Station published a study assessing the genetic variability of gray whales. They determined that the original population might have been as high as 76,000 to 118,000 whales.

How can marine protected areas help wide-ranging cetaceans?

Some managers and marine researchers believe that MPAs may not help in the conservation of most of the wide-ranging and highly mobile whales, dolphins and porpoises. Certainly MPAs cannot cover the entire range of most cetacean populations. Instead, the ideal MPA provides protection for the most important parts of the habitat—where the population feeds, breeds, socializes and raises its calves. To achieve this level of protection may indeed require substantial areas that are then linked together in an MPA network.

For example, the humpback whale in the North Pacific has its breeding grounds protected around Hawaii, as well as feeding areas at the other end of its migration in parts of Alaska and Russia. In the North Atlantic, the humpback feeding areas around Stellwagen Bank National Marine Sanctuary off the northeast United States are linked to the Marine Mammal Sanctuary of the Dominican Republic and the Agoa Sanctuary around the French Caribbean islands of Martinique and Guadeloupe. In both of these humpback whale MPA networks, multiple breeding and feeding grounds have been linked as "sister sanctuaries" to provide protection for the humpback whales, as well as to foster the exchange of scientific and management information about the population.

Often, however, only fragments of a whale population's habitat can be protected. Even so, that protection serves as a valuable starting point for the eventual creation of MPA networks. It offers a benefit in terms of public education and awareness of cetaceans and keeps them on national and international conservation agendas. Additionally, it encourages research and monitoring of cetaceans and their ecosystems.

MPAs as a protection measure are only a few decades old, and most of those for cetaceans have been created in the past decade. Some 600 areas protect important whale, dolphin and porpoise habitat out of a total of 20,000 MPAs in the world ocean for all marine species and habitats. The 20,000 MPAs cover more than 5 percent of the surface of the ocean. At the 2010 Convention on Biological Diversity meeting in Nagoya, Japan, the world's countries agreed that 10 percent of the ocean's surface should be protected by 2020, but many scientific and conservation bodies agree that the minimal level of protection should be at least 30 percent of the ocean. The biggest gap is on the high seas, the international waters that make up more than half of the surface of the ocean, where less than 1 percent has been protected.

Time-consuming and costly to set up, manage and monitor, MPAs are only one tool in the toolbox of conservation solutions. Management plans require engagement from the entire stakeholder community, and management bodies are then needed to implement the plans. Other tools include the re-routing of ships to avoid ship strike and increasing noise levels in areas of known whale abundance, and the banning of set or gillnets in areas where cetaceans are accidentally killed in nets—called "bycatch." In many cases, these other conservation tools will more directly address the problem either used alone or in combination with an MPA.

How can we determine the location of Important Marine Mammal Areas (IMMAs) across the oceans?

A conservation tool called Important Marine Mammal Areas, or IMMAs, is being used to identify the habitats of cetaceans and other marine mammals across the breadth of the ocean and to lay down the best science and expert analysis showing the location of areas that they need for their survival. IMMAs are being developed by the IUCN Marine Mammal Protected Areas Task Force, in partnership with the International Committee on Marine Mammal Protected Areas, Whale and Dolphin Conservation and the Tethys Research Institute.

An Indo-Pacific bottlenose dolphin named Bronny shares life with the pelicans in the Adelaide Dolphin Sanctuary. The river and estuary running through the city of Adelaide, South Australia, was long polluted. It was cleaned up over several decades to make a home for dolphins and other wildlife. The Adelaide Dolphin Sanctuary serves as a prime example of what a group of researchers and citizen scientists can achieve with hard work and dedication.

This work will lead to many more protected area proposals as well as be useful for marine spatial planning by identifying areas where whales need protection because they are getting hit by ships or being caught accidentally in nets from bycatch or where ship or other noise may interrupt their ability to feed, socialize, rest or breed. IMMAs will be a check on how well or how poorly we are protecting whales, dolphins and porpoises. They will also provide baseline-monitoring information in specific areas useful for evaluating climate change in the future.

Should Cetaceans Have Rights?

Are cetacean research subjects meant to be probed, tested, jabbed or worse—forced to spend their lives in a tank? Are they performers in marine parks and aquaria, here for our amusement? Meat if we are hungry? Or do they exist with a purpose of their own, with the right to live their own lives, to have safe and clean habitat and to swim wild and free in the ocean?

If you had asked these questions two hundred years ago, you would get a very different answer than we've heard in the past few decades. If you asked the question in the stands at SeaWorld, the answer would differ from exit interviews from the film *Blackfish*. The answers would differ in Japan, Norway, the United States and Australia.

Common bottlenose dolphins play in their protected habitat in Scotland's Moray Firth Special Area of Conservation.

Even within the research world, some scientists see conservation science as largely independent of welfare concerns for individual animals. In the 1950s and 1960s, scientists eager to learn about whale anatomy and diet were happy to look at carcasses pulled aboard whaling ships or brought into whaling stations, while some studied whale social structure through carcasses. Back then, it was possible to be a good scientist and even a conservationist and yet not be too concerned about individual animals.

Things change. Whaling persists but at a much reduced level. After decades of watching cetaceans up close in the field, many researchers today are sensitive to the welfare of individuals and populations. The specter of extinction with Endangered large whales, river dolphins and others keeps researchers on their toes. At what point should a scientist risk taking a genetic sample from a small dolphin or porpoise, knowing that a misfire could land in the eye or other sensitive area, or even, albeit rarely, result in death?

The question is a matter not only for scientists, but for society as a whole. Clearly, we must care about and pay attention to individuals in order to save species. Individuals make up social groups, social groups make up populations that are breeding units, and it sometimes doesn't take too many populations to make up a species.

Scientists as well as ethicists and legal scholars have started to consider these questions:
- Do whales, dolphins and porpoises—the cetaceans—have "rights"?
- How far do such rights extend?
- Do cetaceans have the right to live their lives

A Charter for Cetacean Rights

The wording of the following declaration, prepared by a group of cetacean scientists and conservationists representing various groups and nationalities, was agreed to on May 22, 2010, in Helsinki, Finland.

Based on the principle of the equal treatment of all persons;

Recognizing that scientific research gives us deeper insights into the complexities of cetacean minds, societies and cultures;

Noting that the progressive development of international law manifests an entitlement to life by cetaceans;

We affirm that all cetaceans as persons have the right to life, liberty and wellbeing.

We conclude that:

Every individual cetacean has the right to life.

No cetaceans should be held in captivity or servitude; be subject to cruel treatment; or be removed from their natural environment.

All cetaceans have the right to freedom of movement and residence within their natural environment.

No cetacean is the property of any State, corporation, human group or individual.

Cetaceans have the right to the protection of their natural environment.

Cetaceans have the right not to be subject to the disruption of their cultures.

The rights, freedoms and norms set forth in this Declaration should be protected under international and domestic law.

Cetaceans are entitled to an international order in which these rights, freedoms and norms can be fully realized.

No State, corporation, human group or individual should engage in any activity that undermines these rights, freedoms and norms.

Nothing in this Declaration shall prevent a State from enacting stricter provisions for the protection of cetacean rights.

To find out more and to sign the declaration and join a global call to have rights formally declared for whales, dolphins and porpoises, go to www.cetaceanrights.org.

without interference from humans? Do they have any rights when it comes to being hunted, captured, made to perform tricks in aquaria or followed by people in boats who want to study or watch and interact with them?

• If cetaceans have rights, should these rights be recognized and enshrined in human laws?

• Do cetaceans have the right to a home—a safe, clean, protected habitat—in the sea?

The issue of animal rights has been widely debated with reference to primates, such as the laboratory chimpanzees with whom humans share nearly 99 percent of their genes. Of course, there is also the question of where to draw the line with animal rights. Do malaria-carrying mosquitoes deserve some rights?

In 2010, a group of farsighted biologists, ethicists and conservationists came together in Helsinki to propose a cetacean bill of rights. Since then, these ideas have been discussed and debated. More and more people today contend that part of our human responsibility of care for Earth is to ensure the free, wild lives of cetaceans, as well as to protect their wild habitats.

Appendix

The Cetaceans

THE CETACEANS

Whales, Dolphins and Porpoises

This chapter features illustrations and fact boxes for the 90 recognized species of cetaceans. These include the four families and 14 species of the baleen whales, the Mysticeti, consisting of the four species of right and bowhead whales (family Balaenidae), the pygmy right whale (family Neobalaenidae), the eight species of rorquals (family Balaenopteridae) and the gray whale (family Eschrichtiidae). The toothed whales, the Odontoceti, comprise 10 families and 76 species. The large-sized toothed whales comprise 27 species and four families. By far the largest is the sperm whale (family Physeteridae). The other members include the pygmy and dwarf sperm whales (family Kogiidae), the beluga and narwhal (family Monodontidae) and the 22 beaked whales (family Ziphiidae). The medium-sized toothed whales comprise the 38 species of oceanic dolphins (family Delphinidae). The small-sized toothed whales comprise the four species of river dolphins (families Platanistidae, Iniidae, Pontoporiidae and Lipotidae) and the seven species of porpoises (family Phocoenidae). Note that in baleen whales, the females are typically larger than the males, while in the toothed whales, males are typically larger than the females. The measurements provided generally represent the greatest lengths and weights recorded for each species.

Previous spread: A massive blue whale off Baja California, Mexico, moves just below the surface as it prepares to feed, then it's back to the deep, facing page.

Calf

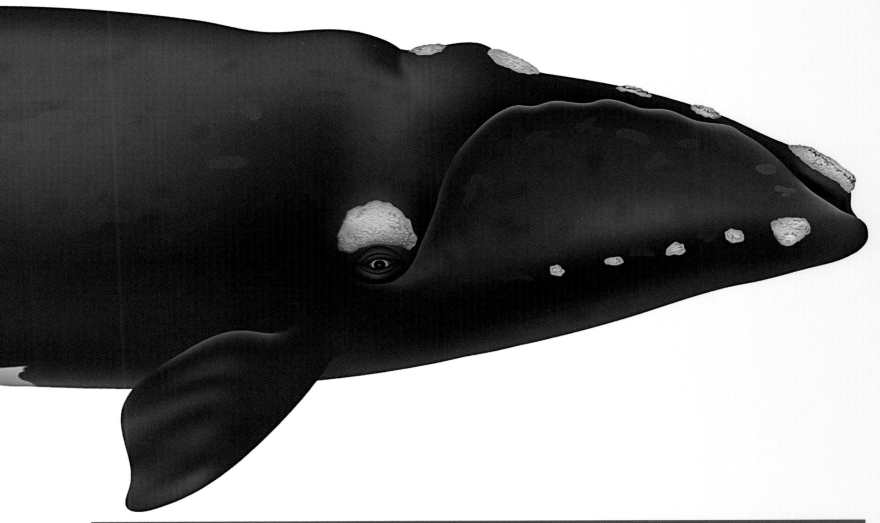

North Atlantic Right Whale

Eubalaena glacialis
Also known as right whale, *baleine de Biscaye, baleine noire, ballena franca*

Length
Males: Up to 59 ft 1 in (18 m)
Females: Up to 59 ft 1 in (18 m)
Calves at birth: 13 ft 1 in–14 ft 9 in (4.0–4.5 m)

Adult weight: Up to 99 tons (90,000 kg)

Habitat and range: Coastal western North Atlantic, with known breeding grounds off northern Florida and Georgia (U.S.) and main feeding grounds in the northwest Atlantic off New England and eastern Canada to Greenland.

Diet: Mainly calanoid copepods; also krill, small invertebrates.

Social notes: Usually alone or in pairs, sometimes up to 12 or more especially seen together in feeding, calving and breeding areas.

Conservation status: The IUCN Red List status is Endangered; only 500 individuals remain, mostly from the western North Atlantic population. Before whaling, there was a substantial eastern North Atlantic population along the coasts of Europe and North Africa, but their numbers today are close to zero.

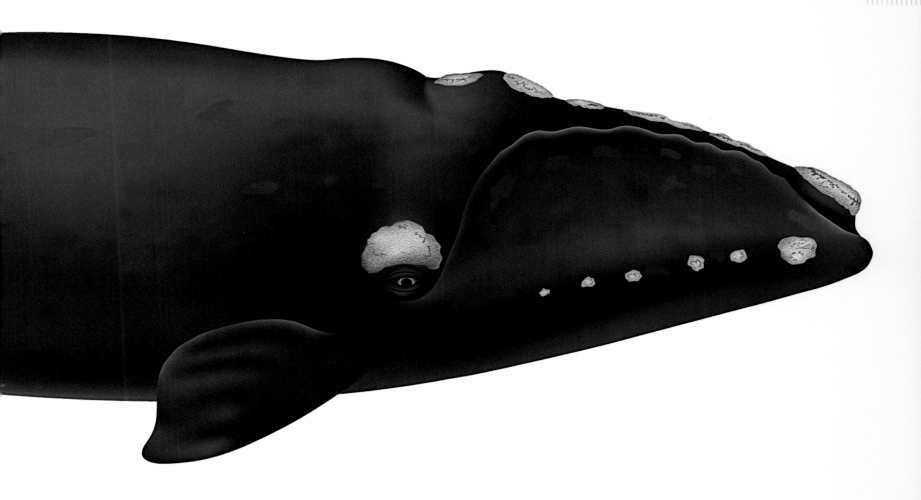

North Pacific Right Whale

Eubalaena japonica
Also known as right whale, *baleine franche du Pacifique, ballena franca del Pacífico*

Length
Males: Up to 62 ft 4 in (19 m)
Females: Up to 62 ft 4 in (19 m)
Calves at birth: 13 ft 1 in–16 ft 5 in
(4–5 m)

Adult weight: Up to 99 tons
(90,000 kg)

Habitat and range: Coastal and offshore waters of the North Pacific, possibly with separate eastern and western populations. Last remaining strongholds for this species may be in the southeastern Bering Sea off Alaska and in Russian waters, especially in the region of the Okhotsk Sea.

Diet: Mainly calanoid copepods; also krill and small invertebrates.

Social notes: Usually alone or in pairs; larger assemblages may form in breeding or feeding areas.

Conservation status: The IUCN Red List status is Endangered with rough estimates of only 500 individuals remaining, mostly in the western North Pacific.

Calf

Southern Right Whale

Eubalaena australis
Also known as right whale, *baleine australe, ballena franca, ballena franca austral*

Length
Males: Up to 55 ft 9 in (17 m)
Females: Up to 55 ft 9 in (17 m)
Calves at birth: 13 ft 1 in–16 ft 5 in (4–5 m)

Adult weight: Up to 88 tons (80,000 kg)

Habitat and range: Coastal circumpolar waters in the Southern Hemisphere with breeding areas, often nearshore, off southern South America, South Africa, Australia, New Zealand and various temperate and subantarctic islands; feeding areas located farther south as far as Antarctic waters.

Diet: Copepods and krill.

Social notes: With larger numbers than the North Atlantic and North Pacific right whales, southern rights are often seen in informal "groups" of dozens, spread out over the breeding ground.

Conservation status: The IUCN Red List status is Least Concern, but the Chile-Peru population is considered Critically Endangered.

Calf

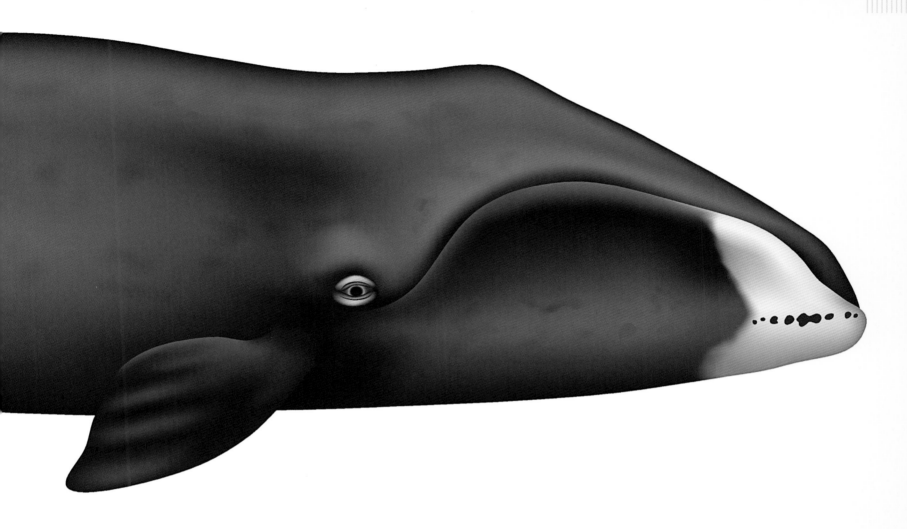

Bowhead Whale

Balaena mysticetus
Also known as bowhead, Greenland right whale, Arctic right whale, *baleine de grande baie, baleine du Groenland, ballena boreal, ballena de Groenlandia*

Length
Males: Up to 59 ft 1 in (18 m)
Females: Up to 65 ft 7 in (20 m)
Calves at birth: 13 ft 1 in–14 ft 9 in
(4.0–4.5 m)

Adult weight (estimated): 99 tons
(90,000 kg)

Habitat and range: Mainly
Arctic and subarctic range with
distribution in summer and winter
related to the advance and retreat of
pack ice.

Diet: Krill, copepods, mysids and
other invertebrates.

Social notes: Sometimes seen in
informal associations of twos or
threes, while larger aggregations are
found on feeding grounds and when
traveling on migration.

Conservation status: The IUCN Red
List status overall is Least Concern,
but the Okhotsk Sea population is
considered Endangered, and the
Svalbard-Barents Sea (Spitsbergen)
population is Critically Endangered.

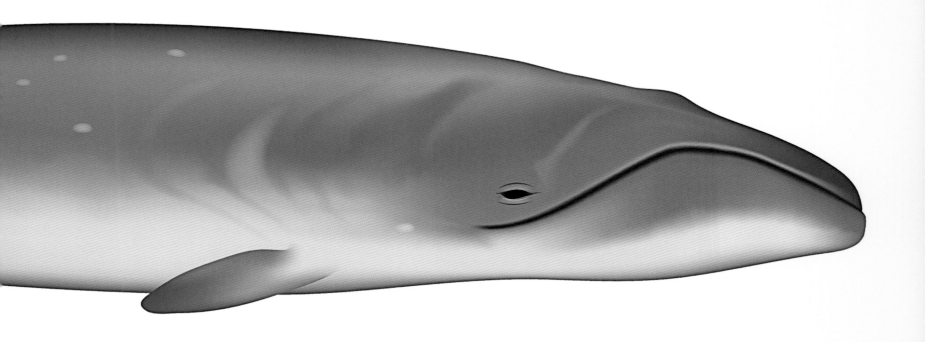

Pygmy Right Whale

Caperea marginata
Also known as pygmy right, *baleine pygmée, ballena franca pigmea*

Length
Males: Up to 20 ft (6.1 m)
Females: Up to 21 ft 4 in (6.5 m)
Calves at birth: 6 ft 7 in (2 m)

Adult weight: At least 7,495 lb
(3,400 kg)

Habitat and range: Coastal and offshore circumpolar waters of the Southern Hemisphere (based on few records).

Diet: Copepods, small euphausiids (krill) and other small invertebrates.

Social notes: Singles, casual associations of two to 14 individuals; sometimes larger feeding assemblages are witnessed.

Conservation status: The IUCN Red List status is Data Deficient.

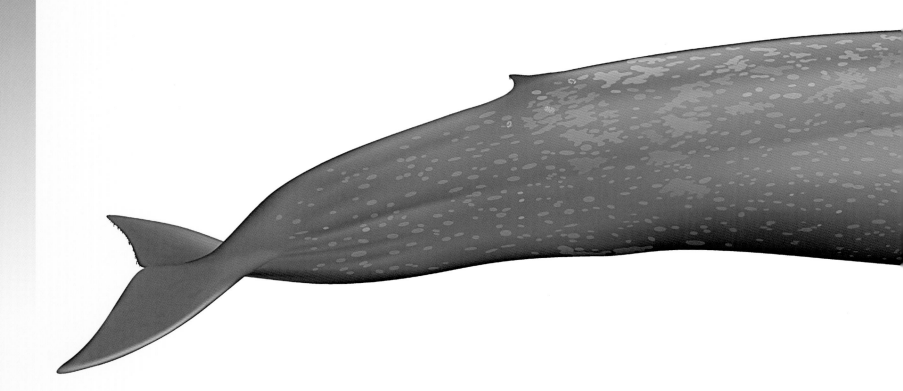

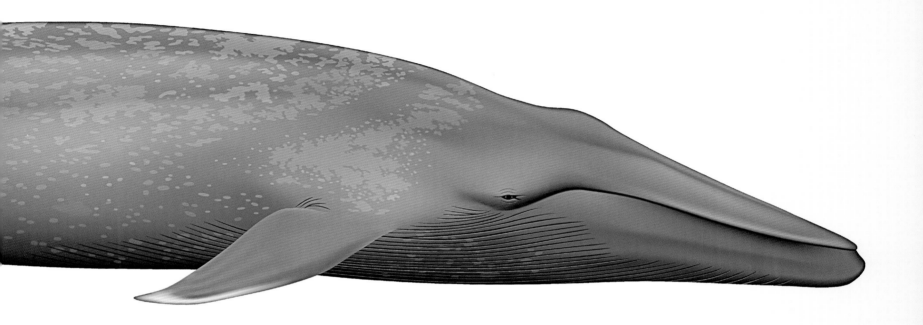

Blue Whale

Balaenoptera musculus
Also known as Sibbald's rorqual, sulphur-bottom whale, pygmy blue whale (separate population), *rorqual bleu, baleine bleu, baleinoptère bleue, ballena azul, rorcual azul*

Length
Males: Less than 108 ft 2 in (33 m)
Females: Up to 108 ft 2 in (33 m)
Calves at birth: 23–26 ft 3 in (7–8 m)

Adult weight: Typically 79–149 tons (72,000–135,000 kg) but up to 198 tons (180,000 kg)

Habitat and range: World ocean, except for the Arctic, including the Bering Sea. Habitat is mainly deeper waters but coastal in some areas where deep waters and krill concentrations are close to shore (e.g., the St. Lawrence River and Gulf in Canada).

Diet: Krill (euphausiids) of various species, depending on location.

Social notes: Seen alone or in pairs; aerial surveys often reveal larger aggregations widely spread out on feeding grounds. Associations, except mothers and calves, are thought to be mostly short-term.

Conservation status: The IUCN Red List status is Endangered, but the Antarctic blue whale population is rated Critically Endangered.

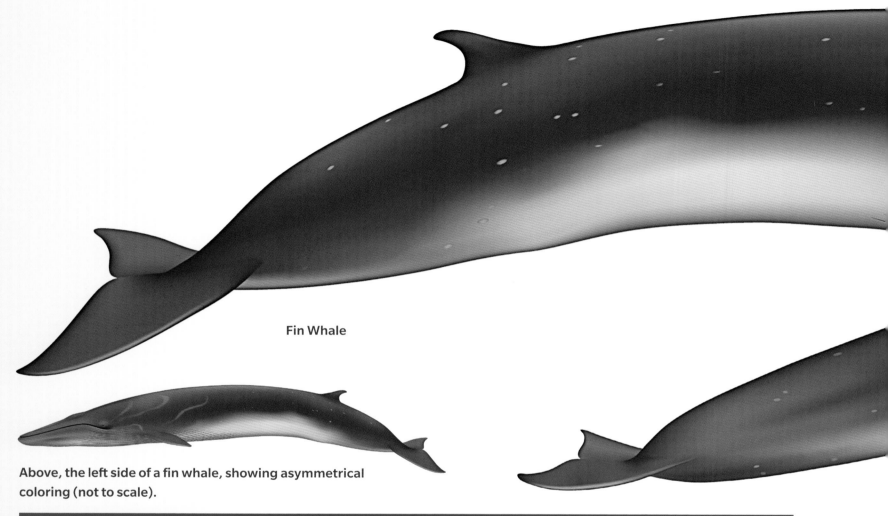

Fin Whale

Above, the left side of a fin whale, showing asymmetrical coloring (not to scale).

Fin Whale

Balaenoptera physalus
Also known as finback, finner, common rorqual, razorback, *baleine fin, rorqual commun, ballena de aleta, rorcual común*

Length
Males: Up to 82 ft (25 m)
Females: Up to 88 ft 7 in (27 m)
Calves at birth: 19 ft 8 in–
21 ft 4 in (6.0–6.5 m)

Adult weight: Typically less than
99 tons (90,000 kg) but occasionally
up to 132 tons (120,000 kg).

Habitat and range: Mainly offshore
world ocean, although they can
be coastal when food supplies are
present. Rarely seen in tropical
waters; polar distribution but rarely
to the high Arctic and Antarctic.

Diet: Krill (euphausiids), copepods
and other small invertebrates;
capelin, herring, sand lance and
other schooling fishes.

Social notes: Often feed alone in
some areas; in food-rich areas,
there may be loose associations of
two to seven or more whales. They
sometimes travel and feed with
blue, minke and humpback whales.
Blue-fin hybrids occasionally occur.

Conservation status: The IUCN
Red List status is Endangered, while
the fin whale population in the
Mediterranean is rated Vulnerable.

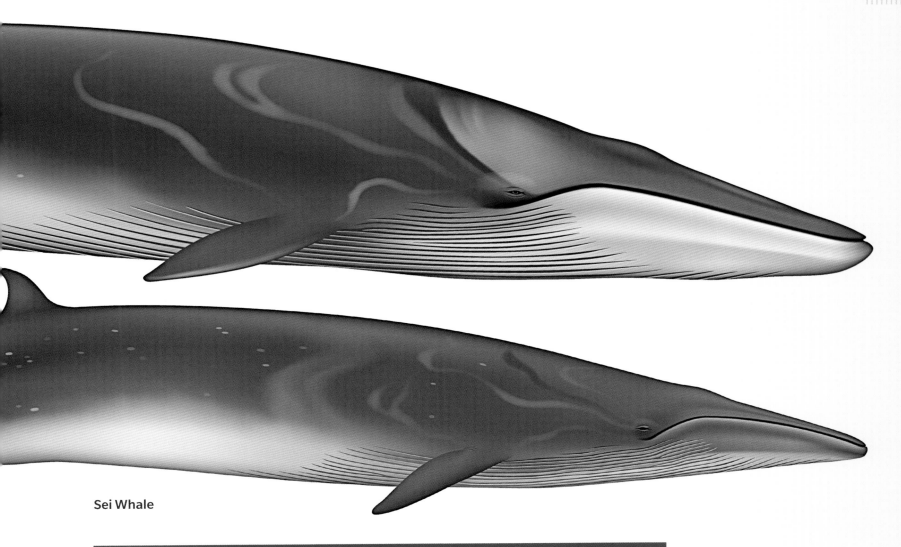

Sei Whale

Sei Whale

Balaenoptera borealis
Also known as pollack whale, coalfish whale, Rudolphi's rorqual, *rorqual sei, rorqual de Rudolphi, rorqual boreal,*
ballena boba, ballena sei, rorcual de Rudolphi

Length
Males: Less than 59 ft 1 in (18 m)
Females: Up to 59 ft 1 in (18 m)
Calves at birth: 14 ft 9 in–
15 ft 9 in (4.5–4.8 m)

Adult weight: Up to 49 tons
(45,000 kg)

Habitat and range: Mainly offshore
waters of the world ocean, except
for northern Indian Ocean and the
high Arctic and Antarctic.

Diet: Skim feeding: copepods, other
small invertebrates; lunge feeding:
sardines, anchovies, cephalopods,
krill.

Social notes: Travel and feed alone
or occasionally in loose associations
with other seis.

Conservation status: The IUCN Red
List status is Endangered.

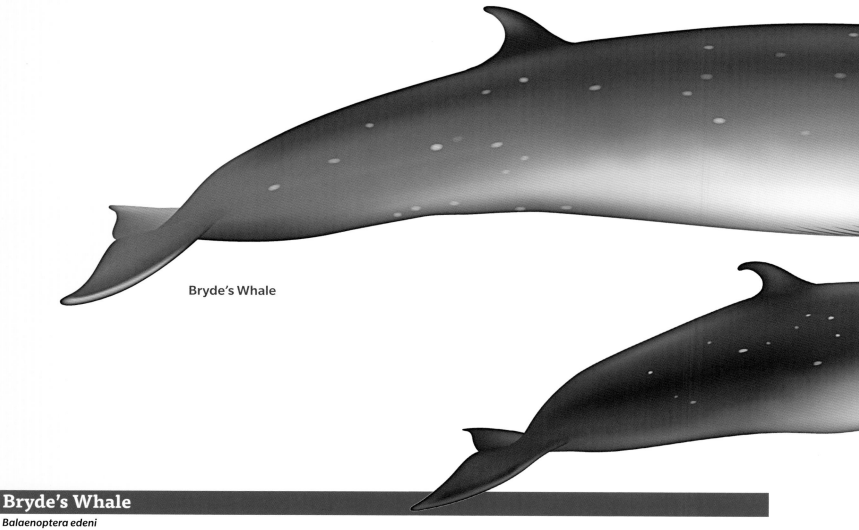

Bryde's Whale

Bryde's Whale

Balaenoptera edeni
Also called *Balaenoptera brydei*. The taxonomy is unresolved, but in 2015, some researchers argued that there are two separate species, and they should be recognized as such. Also known as Eden's whale, pygmy Bryde's whale, tropical whale, *rorqual de Bryde, rorqual d'Eden, rorqual tropical, ballena de Bryde*

Length
Males: Up to 49 ft 3 in (15 m)
Females: Up to 54 ft 2 in (16.5 m)
Calves at birth: 13 ft 1 in (4 m)

Adult weight: Up to 44 tons
(40,000 kg)

Habitat and range: Tropical and subtropical waters, coastal to offshore, in the Atlantic, Pacific and Indian oceans.

Diet: Schooling fishes, such as sardine, herring and anchovy, as well as krill and squid.

Social notes: Usually travel alone or in pairs but sometimes seen on feeding grounds in temporal associations of up to 20 individuals.

Conservation status: The IUCN Red List status is Data Deficient. Bryde's whales may be one or several species. The main Bryde's whale has a worldwide tropical and subtropical distribution. In addition, there are smaller forms that tend to be more coastal in distribution. These may be separate populations, subspecies or even species.

Above, the left side of an Omura's whale, showing asymmetrical coloring (not to scale).

Omura's Whale

Omura's Whale

Balaenoptera omurai
Also known as *rorqual de Omura, ballena de Omura*

Length
Males: Up to 39 ft 5 in (12 m)
Females: Up to 39 ft 5 in (12 m)
Calves at birth: 11 ft 6 in–13 ft 1 in (3.5–4.0 m)

Adult weight: Up to 22 tons (20,000 kg)

Habitat and range: Coastal and nearshore tropical and subtropical waters of the eastern Indian Ocean and western Pacific; some sightings in other areas.

Diet: Possibly schooling fishes.

Social notes: Usually travel alone or in pairs with larger associations during feeding.

Conservation status: The IUCN Red List status is Data Deficient.

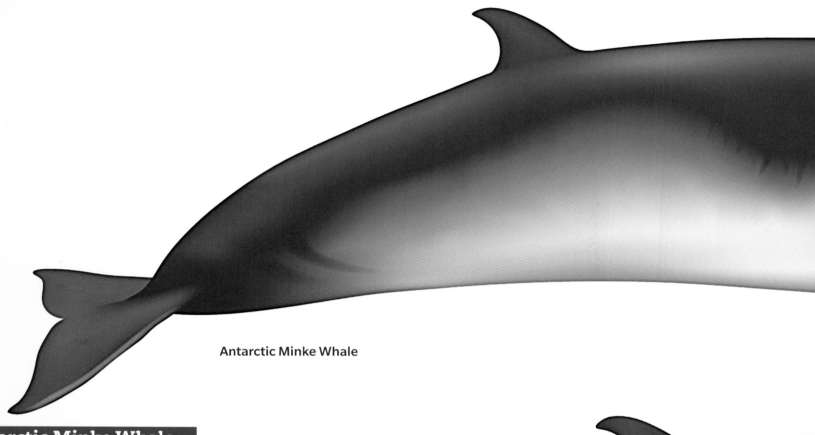

Antarctic Minke Whale

Antarctic Minke Whale

Balaenoptera bonaerensis
Also known as minke, *petit rorqual, rorcual enano, ballena minke, rorcual menor*

Length
Males: Up to 35 ft 1 in (10.7 m)
Females: Up to 35 ft 1 in (10.7 m)
Calves at birth: 9 ft 2 in (2.8 m)

Adult weight: Up to 10 tons (9,100 kg)

Habitat and range: Coastal to offshore waters of the Southern Hemisphere.

Diet: Krill and small schooling fishes.

Social notes: Alone or in informal groups of two or three with occasional larger associations during feeding.

Conservation status: The IUCN Red List status is Data Deficient.

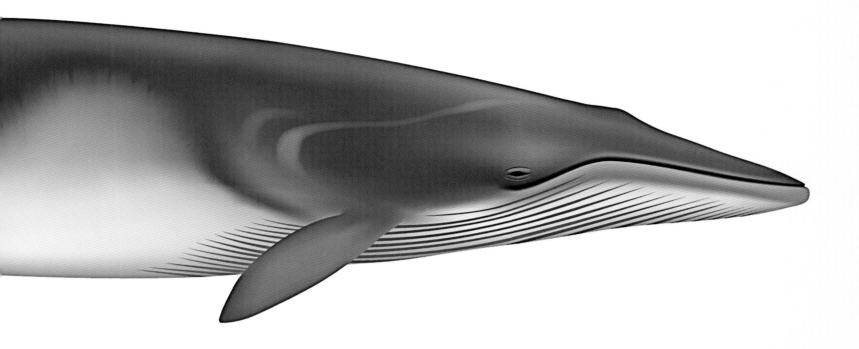

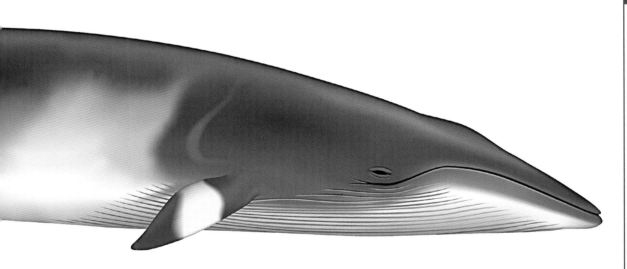

Common Minke Whale

Common Minke Whale

Balaenoptera acutorostrata
**Also known as minke, dwarf minke whale
(separate population), little piked whale,
lesser rorqual, *petit rorqual, rorcual enano,
ballena minke, rorcual menor***

Length
Males: Up to 26 ft 3 in (8 m)
Females: Up to 28 ft 10 in (8.8 m)
Calves at birth: 6 ft 7 in–9 ft 2 in (2.0–2.8 m)

Adult weight: Up to 10 tons (9,200 kg)

Habitat and range: World ocean, except
southern ocean, coastal to pelagic.

Diet: Schooling fishes such as mackerel,
capelin, sand lance, sand eel, coal fish,
whiting, lanternfish, as well as some larger fish
species such as haddock and dogfish.

Social notes: Alone or in informal groups of
two or three; also larger associations.

Conservation status: The IUCN Red List status
is Least Concern.

Humpback Whale

Megaptera novaeangliae
Also known as humpback, *baleine à bosse*, *mégaptère*, *jubarte*, *ballena jorobada*

Length
Males: Up to 52 ft 9 in (16 m)
Females: Up to 55 ft 9 in (17 m)
Calves at birth: 14 ft 1 in (4.3 m)

Adult weight: At least 44 tons (40,000 kg)

Habitat and range: World ocean from nearshore to offshore. Cold-water feeding areas are circumpolar extending into the high latitudes, while breeding areas are concentrated around shallow subtropical island archipelagos and continents.

Diet: Krill, schooling fishes including herring, capelin, sand lance, mackerel, sardines.

Social notes: Usually one to three individuals, depending on behavior. Cooperative bubble feeding might include more than 20 individuals, with bubble structures varying regionally (bubble nets and clouds in the North Atlantic; bubble nets only in the North Pacific). Competitive groups on the breeding grounds can involve anywhere from two to 20 or more males competing for a single female.

Conservation status: The IUCN Red List status is Least Concern and increasing in many areas, but they have yet to return to original numbers. However, two populations of humpbacks in the Arabian Sea and in Oceania are still rated Endangered.

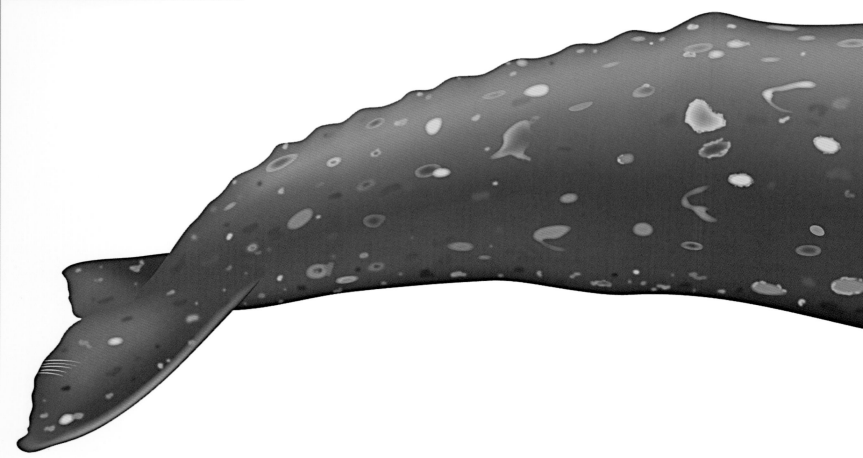

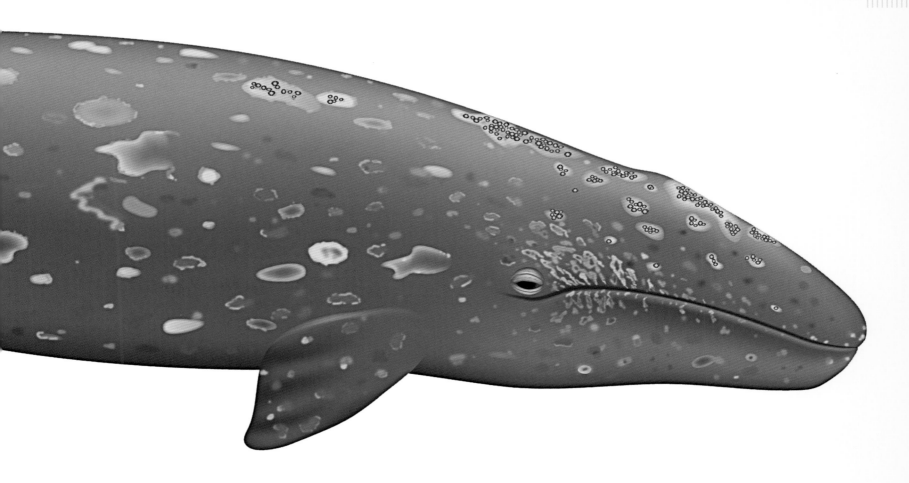

Gray Whale

Eschrichtius robustus
Also known as grey whale, *baleine grise*, *ballena gris*

Length
Males: Less than 49 ft 3 in (15 m)
Females: Up to 49 ft 3 in (15 m)
Calves at birth: 15 ft 1 in–16 ft 1 in
(4.6–4.9 m)

Adult weight: Up to 49.6 tons
(45,000 kg)

Habitat and range: Coastal, mainly eastern North Pacific from Mexico (breeding grounds) to Alaska and Siberian Russia and southernmost Arctic Ocean; a small western North Pacific gray whale population is found off Sakhalin Island, Russia, but breeding ground is unknown.

Diet: Mysids, amphipods and polychaete tubeworms, various invertebrates, including larvae and fish eggs.

Social notes: Usually in pairs (often mother and calf), sometimes threes; occasionally larger short-term groups on migration.

Conservation status: The IUCN Red List status is Least Concern, but the western population is rated Critically Endangered.

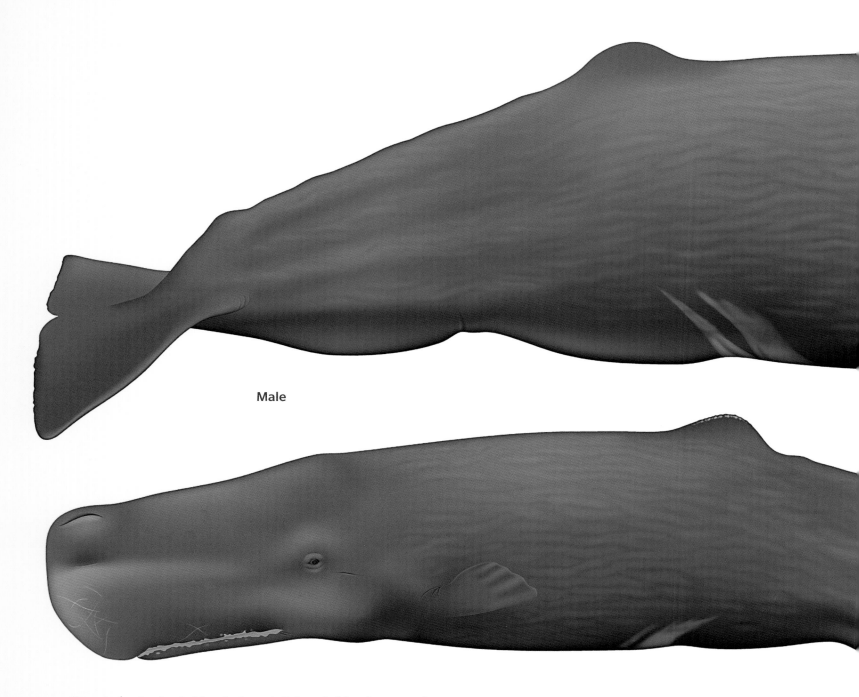

Male

Female (note single blowhole on left-hand side of rostrum)

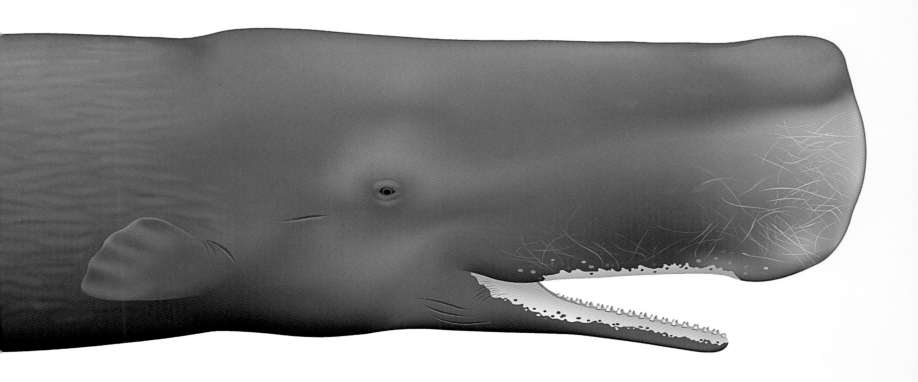

Sperm Whale

Physeter macrocephalus
Also known as *cachalot, cachalote*

Length
Males: Up to 63 ft (19.2 m)
Females: Up to 41 ft (12.5 m)
Calves at birth: 11 ft 6 in–14 ft 9 in (3.5–4.5 m)

Adult weight: Up to 69 tons (57,000 kg)

Habitat and range: World ocean, mainly from edge of continental shelf to deep trenches; deep water when close to shore.

Diet: Giant squid and other cephalopods, including octopuses, deep-sea fishes such as redfishes and lumpsuckers.

Social notes: Groups of 20 to 30 females and young, sometimes more than 50. Males can be single or in small groups.

Conservation status: The IUCN Red List status is Vulnerable. Sperm whale populations in most oceans, except notably southeast Pacific, are rebounding from intensive whaling but have not yet recovered.

Pygmy Sperm Whale

Dwarf Sperm Whale

Pygmy Sperm Whale

Kogia breviceps
Also known as *cachalot pygmée, cachalote pigmeo*

Length
Males: Up to 12 ft 6 in (3.8 m)
Females: Up to 12 ft 6 in (3.8 m)
Calves at birth: 3 ft 11 in (1.2 m)

Adult weight: Up to 992 lb (450 kg)

Habitat and range: Offshore (outer continental shelf) waters of the tropical to temperate world ocean; in some areas, in deeper waters than the dwarf sperm whale.

Diet: Deep-water cephalopods, sometimes fish and shrimps.

Social notes: Small groups of up to six individuals.

Conservation status: The IUCN Red List status is Data Deficient.

Dwarf Sperm Whale

Kogia sima
Also known as *cachalote enano, cachalot nain*

Length
Males: Up to 8 ft 10 in (2.7 m)
Females: Up to 8 ft 10 in (2.7 m)
Calves at birth: 3 ft 3 in (1 m)

Adult weight: Up to 560 lb (272 kg)

Habitat and range: Offshore (outer continental shelf) waters of the tropical to warm temperate world ocean, preferring to stay in warmer waters than does the pygmy sperm whale.

Diet: Deep-water cephalopods and other prey.

Social notes: Small groups of up to six individuals, sometimes up to 16.

Conservation status: The IUCN Red List status is Data Deficient.

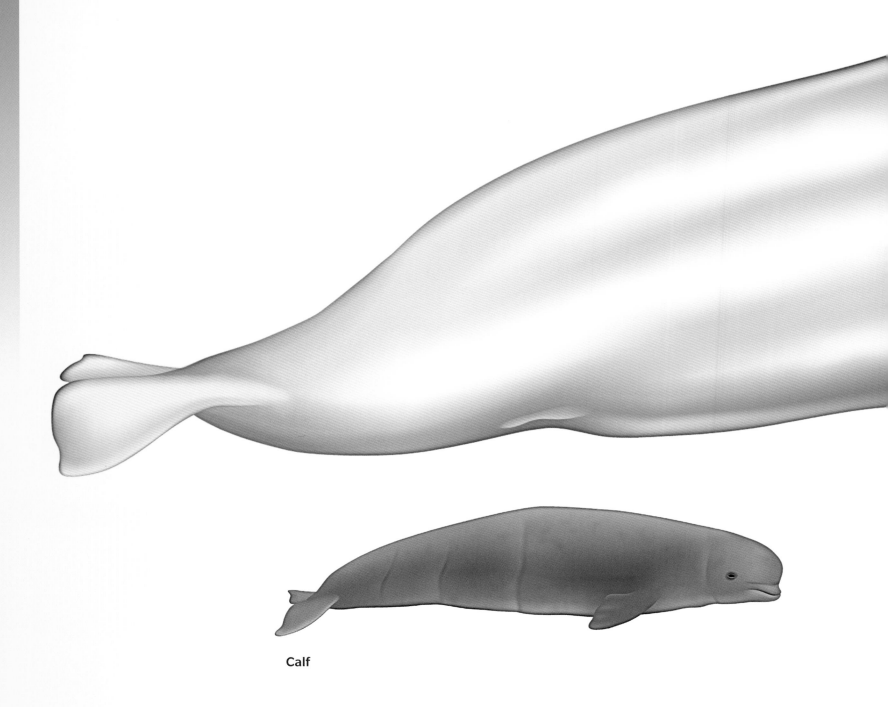

Calf

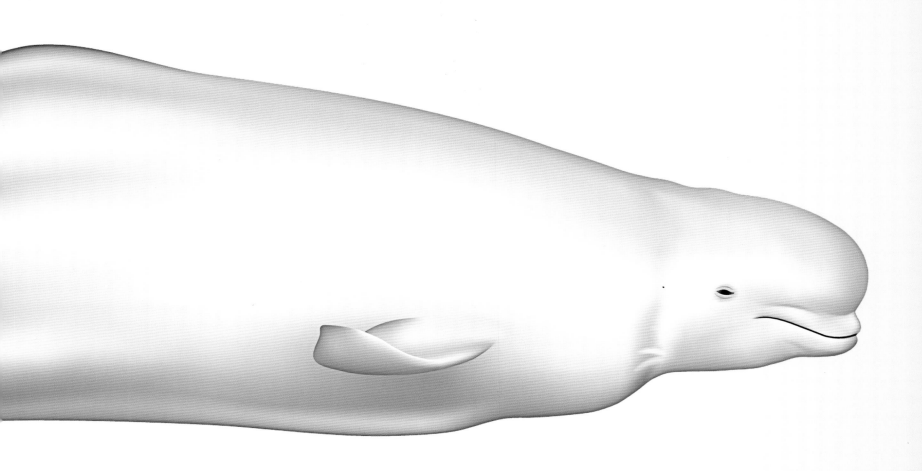

Beluga

Delphinapterus leucas
Also known as white whale, *dauphin blanc, bélouga, delphinaptère blanc, marsouin blanc, ballena blanca*

Length
Males: Up to 18 ft (5.5 m)
Females: Up to 14 ft 1 in (4.3 m)
Calves at birth: 5 ft 3 in (1.6 m)

Adult weight: Up to 3,527 lb
(1,600 kg)

Habitat and range: Shallow to deep
waters of the circumpolar Arctic and
subarctic.

Diet: Main prey items are salmon,
herring and Arctic cod; also squid,
octopus, benthic shrimps and
crabs.

Social notes: Highly social whales
seen in groups of 15 that may
congregate to form migratory
groups of thousands.

Conservation status: The IUCN Red
List status is Near Threatened. The
Cook Inlet population in Alaskan
waters is considered Critically
Endangered. A separate population
in the St. Lawrence River is
recovering slowly from the cocktail
of pollutants in the St. Lawrence
carried from industries based
upstream and in the Great Lakes.

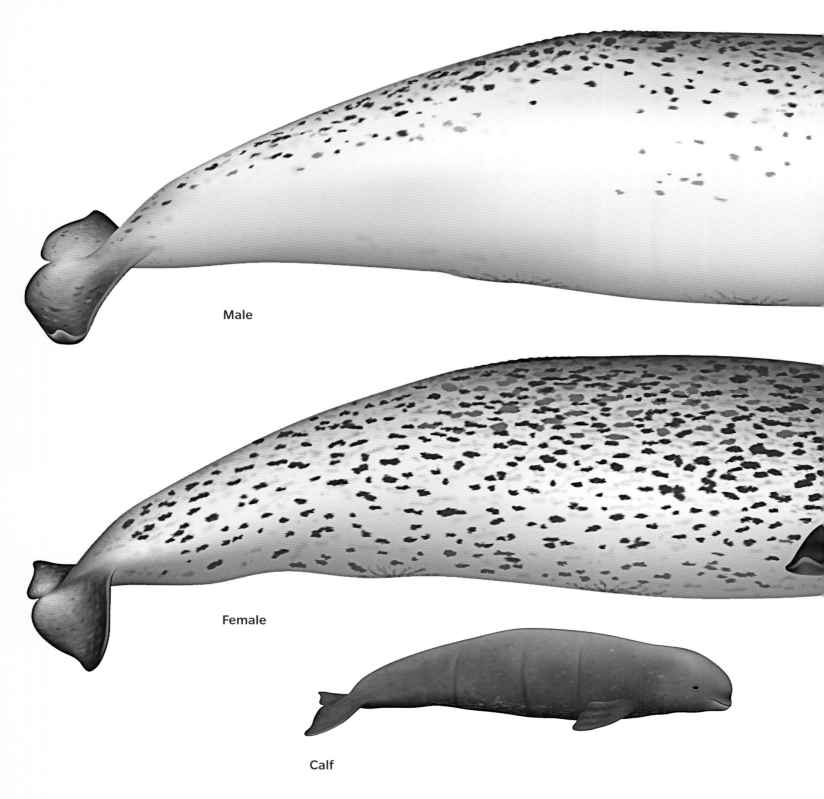

Male

Female

Calf

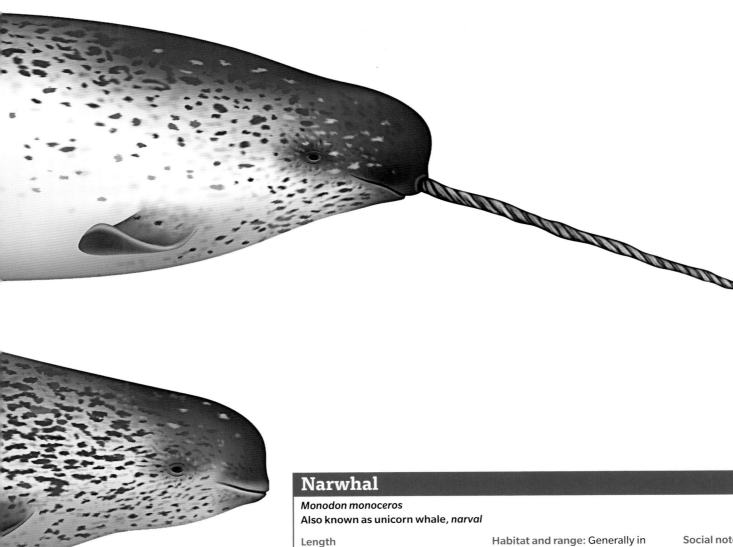

Narwhal

Monodon monoceros
Also known as unicorn whale, *narval*

Length
Males: Up to 15 ft 9 in (4.8 m),
not including tusk(s)
Females: Up to 13 ft 9 in (4.2 m)
Calves at birth: 5 ft 3 in (1.6 m)

Adult weight: 2,205–3,527 lb
(1,000–1,600 kg)

Habitat and range: Generally in deeper and higher Arctic waters than belugas, entirely within Arctic Circle and concentrated in the Atlantic portion of the Arctic, mainly from the central Canadian Arctic around to the eastern Russian Arctic.

Diet: Medium to large-sized Arctic fish species, including turbot, Arctic cod and polar cod, plus squid and shrimp

Social notes: Up to 10 individuals in winter with much larger groups of up to thousands in summer, though often spread apart. Males sometimes travel in separate groups from females and calves.

Conservation status: The IUCN Red List status is Near Threatened.

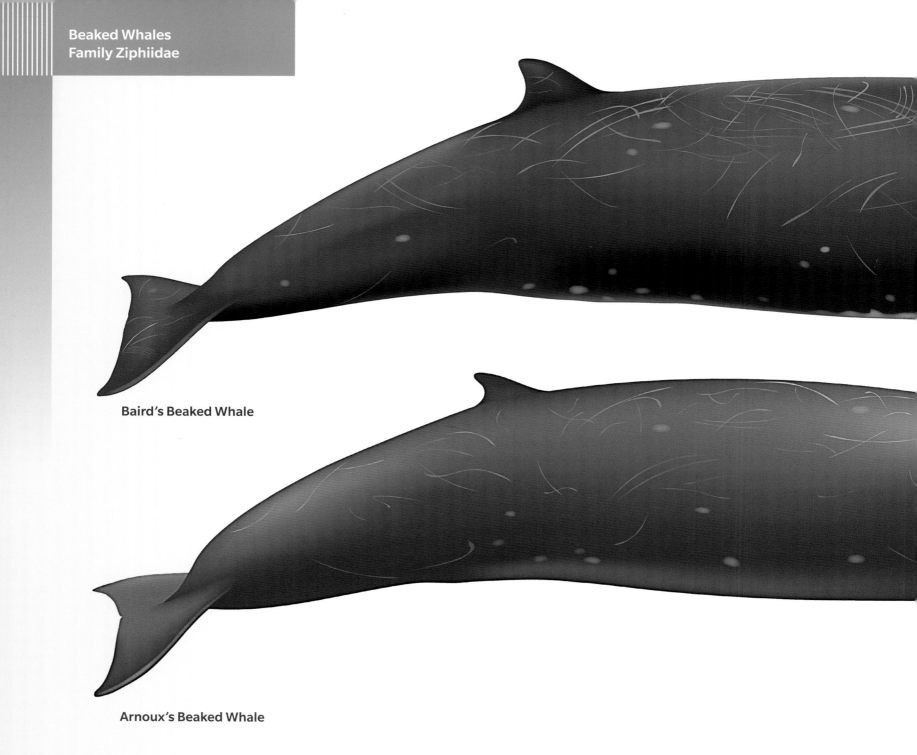

Baird's Beaked Whale

Arnoux's Beaked Whale

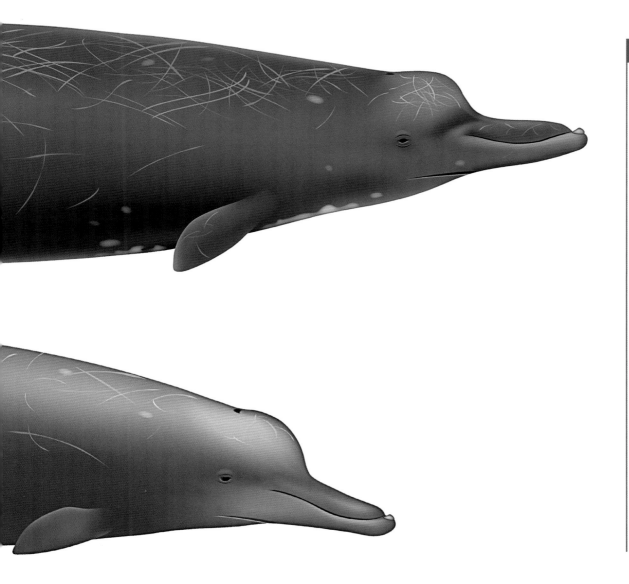

Baird's Beaked Whale

Berardius bairdii
Also known as Pacific giant bottlenose whale, northern four-toothed whale, *baleine à bec de Baird, ballena de pico de Baird, zifio de Baird*

Length
Males: Up to 35 ft 1 in (10.7 m)
Females: Up to 36 ft 5 in (11.1 m)
Calves at birth: 15 ft 1 in (4.6 m)

Adult weight: Up to 13 tons (12,000 kg)

Habitat and range: Deep waters from the edge of the continental shelf in the North Pacific, temperate to subarctic, including Okhotsk and East Sea (Sea of Japan) in the west and down to southern Gulf of California in the eastern Pacific.

Diet: Deep-water gadiform fishes, cephalopods and crustaceans; sometimes mackerel, sardines and other pelagic fishes; in some areas, the diet is mainly fish while in others, cephalopods dominate.

Social notes: Larger group sizes than most beaked whales, ranging from five to 20 individuals, often about eight. Around Bering Island, there is evidence of some associations with older individuals lasting over several years.

Conservation status: The IUCN Red List status is Data Deficient.

Arnoux's Beaked Whale

Berardius arnuxii
Also known as southern giant bottlenose whale, southern four-toothed whale, *béradien d'Arnoux, ballena de pico de Arnoux, ballenato de Arnoux*

Length
Males: Up to 30 ft 6 in (9.3 m)
Females: Up to 30 ft 6 in (9.3 m)
Calves at birth: 13 ft 1 in (4 m)

Adult weight: Unknown

Habitat and range: Circumpolar deep waters in the Southern Hemisphere, extending from the Antarctic ice to warm temperate waters off east and west coasts of Australia, South Africa and South America.

Diet: Squid and perhaps deep-water fishes.

Social notes: Up to 10 individuals form typical groups, with larger aggregations of up to 80.

Conservation status: The IUCN Red List status is Data Deficient.

Northern Bottlenose Whale

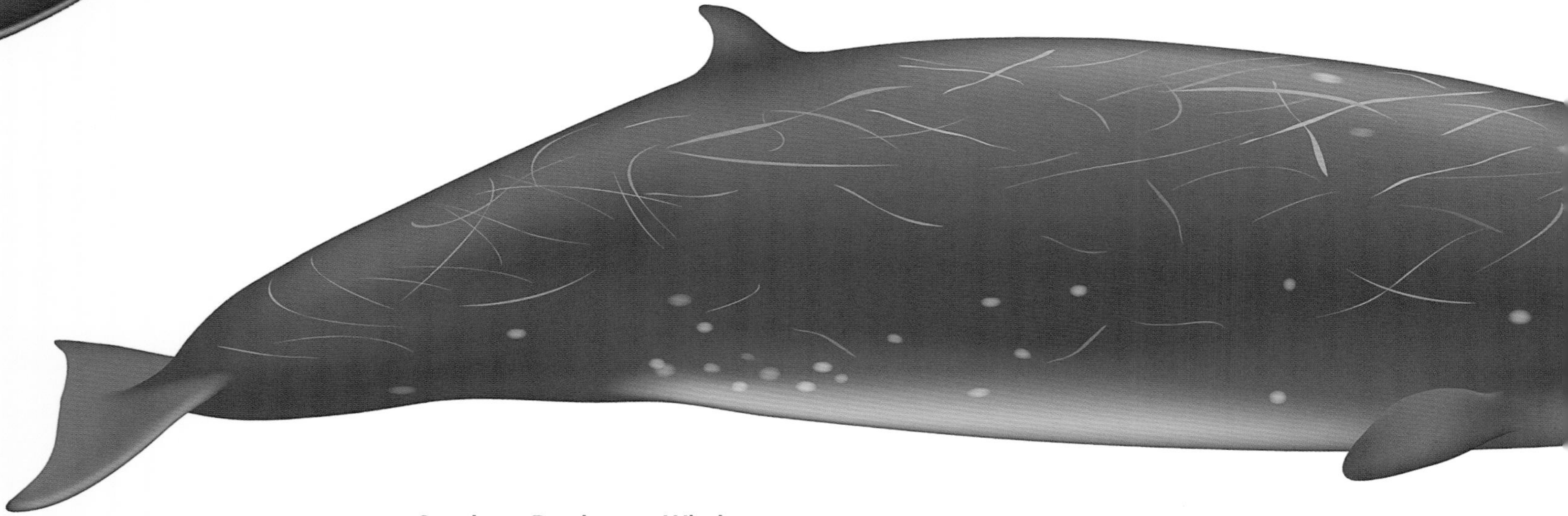

Southern Bottlenose Whale

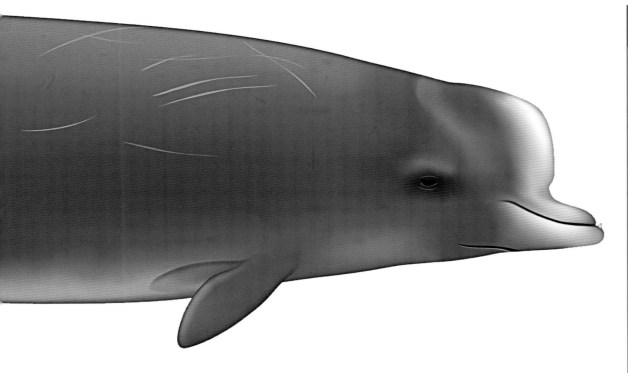

Northern Bottlenose Whale

Hyperoodon ampullatus
Also known as bottlehead, North Atlantic bottlenose whale, *hyperoodon boréal*, *ballena hocico de botella del norte*, *ballena nariz de botella del norte*

Length
Males: Up to 32 ft 10 in (10 m)
Females: Up to 28 ft 3 in (8.6 m)
Calves at birth: 9 ft 10 in–
11 ft 6 in (3.0–3.5 m)

Adult weight: Up to 16,534 lb (7,500 kg)

Habitat and range: Deep offshore waters of the cold temperate to subarctic North Atlantic, mainly off the continental shelf, extending north to the edge of the ice.

Diet: Deep-water squid, especially *Gonatus* squid, plus herring, deep-sea fishes and shrimps.

Social notes: Often seen in groups of four, sometimes up to 20 individuals. Groups sometimes form according to sex and age class.

Conservation status: The IUCN Red List status is Data Deficient.

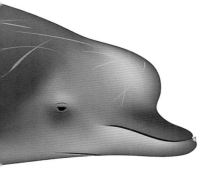

Southern Bottlenose Whale

Hyperoodon planifrons
Also known as bottlenose whale, flat-headed bottlenose whale, *hyperoodon austral*, *ballena nariz de botella del sur*

Length
Males: Up to 23 ft (7 m)
Females: Up to 24 ft 7 in (7.5 m)
Calves at birth: 6 ft 7 in–9 ft 10 in (2–3 m)

Adult weight (estimated): 8,818 lb (4,000 kg)

Habitat and range: Circumpolar deep waters off the continental shelf and usually some distance from the ice in the Southern Hemisphere, extending from the Antarctic to warm temperate waters off east and west coasts of Australia, South Africa and South America.

Diet: Mainly squid, some crustaceans and fish (including Patagonian toothfish).

Social notes: Groups of typically fewer than 10 individuals, sometimes up to 25.

Conservation status: The IUCN Red List status is Least Concern.

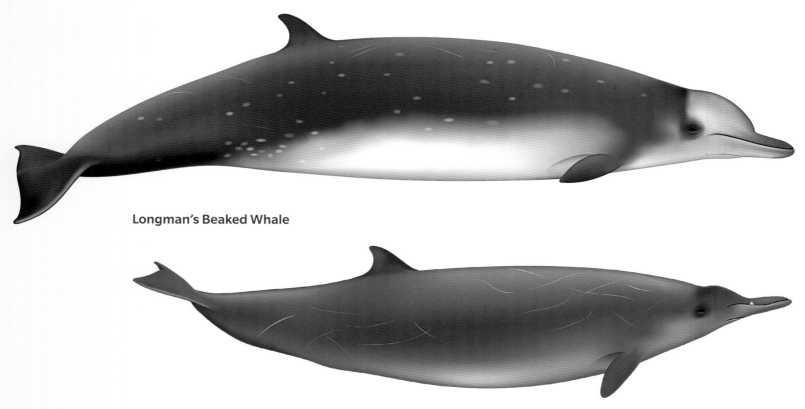

Longman's Beaked Whale

Sowerby's Beaked Whale

Longman's Beaked Whale

Indopacetus pacificus
Also known as tropical bottlenose whale, Indo-Pacific beaked whale,
baleine à bec de Longman, zifio de Longman

Length
Males: Size unknown
Females: Up to 21 ft 4 in (6.5 m)
Calves at birth (estimated): 9 ft 6 in
(2.9 m)

Adult weight: Unknown

Habitat and range: Deep offshore
tropical waters of the Pacific and
Indian oceans.

Diet: Cephalopods.

Social notes: Groups of 10
individuals or more (up to 100)
may form.

Conservation status: The IUCN
Red List status is Data Deficient.

Sowerby's Beaked Whale

Mesoplodon bidens
Also known as North Sea beaked whale, North Atlantic beaked whale,
baleine à bec de Sowerby, zifio de Sowerby, ballena de pico de Sowerby,
mésoplodon de Sowerby

Length
Males: Up to 18 ft (5.5 m)
Females: Up to 16 ft 9 in (5.1 m)
Calves at birth: 7 ft 11 in (2.4 m)

Adult weight: Up to 2,866 lb
(1,300 kg)

Habitat and range: Deep, colder
offshore waters of the North
Atlantic, especially eastern North
Atlantic.

Diet: Squid and small fishes,
including Atlantic cod.

Social notes: Groups of three to
10 individuals, mixed composition.

Conservation status: The IUCN
Red List status is Data Deficient.

Andrews' Beaked Whale

Hubbs' Beaked Whale

Andrews' Beaked Whale

Mesoplodon bowdoini
Also known as splaytooth beaked whale, *baleine à bec de Bowdoin, zifio de Andrews, ballena de pico de Andrew*

Length
Males: Up to 14 ft 5 in (4.4 m)
Females: Up to 14 ft 5 in (4.4 m)
Calves at birth: 7 ft 3 in (2.2 m)

Adult weight: Unknown

Habitat and range: Deep offshore, mostly circumpolar waters of the temperate Southern Hemisphere, based on only a few dozen stranding records, mainly from New Zealand.

Diet: Cephalopods.

Social notes: Unknown, never identified alive at sea.

Conservation status: The IUCN Red List status is Data Deficient.

Hubbs' beaked whale

Mesoplodon carlhubbsi
Also known as arch-beaked whale, mésoplodon de Hubbs, zifio de Hubbs, ballena picuda de Hubbs, baleine à bec de Hubbs

Length
Males: Up to 17 ft 9 in (5.4 m)
Females: Up to 17 ft 9 in (5.4 m)
Calves at birth: 8 ft 2 in (2.5 m)

Adult weight: Over 3,307 lb (1,500 kg)

Habitat and range: Deep offshore temperate waters of the North Pacific, with the few strandings found off Japan and off the west coast of North America from California to British Columbia.

Diet: Squid and some deep-water fishes.

Social notes: Unknown, few sightings in the wild.

Conservation status: The IUCN Red List status is Data Deficient.

Gervais' Beaked Whale

Ginkgo-toothed Beaked Whale

Gervais' Beaked Whale

Mesoplodon europaeus
**Also known as Gulf Stream beaked whale, *mésoplodon de Gervais,
baleine à bec de Gervais, zifio de Gervais, ballena de pico de Gervais***

Length
Males: Up to 15 ft 1 in (4.6 m)
Females: Up to 15 ft 9 in (4.8 m)
Calves at birth: 6 ft 11 in (2.1 m)

Adult weight: At least 2,646 lb
(1,200 kg)

Habitat and range: Deep
offshore waters of the tropical
and temperate North and South
Atlantic.

Diet: Mainly squid.

Social notes: Unknown, rarely seen
at sea.

Conservation status: The IUCN
Red List status is Data Deficient.

Ginkgo-toothed Beaked Whale

Mesoplodon ginkgodens
**Also known as *baleine à bec de Nishiwaki, zifio Japonés, ballena picuda de
dientes de ginkgo, zifio de Nishiwaki***

Length
Males: Up to 17 ft 5 in (5.3 m)
Females: Up to 17 ft 5 in (5.3 m)
Calves at birth: 6 ft 7 in–
8 ft 2 in (2.0–2.5 m)

Adult weight: Unknown

Habitat and range: Deep offshore
waters of the temperate and
tropical Pacific, based on a few
dozen strandings.

Diet: Squid.

Social notes: Unknown, no
documented sightings at sea.

Conservation status: The IUCN
Red List status is Data Deficient.

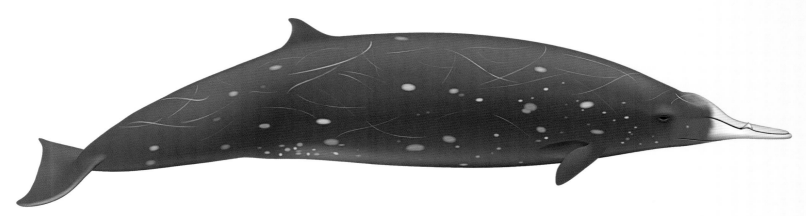

Gray's Beaked Whale

Hector's Beaked Whale

Gray's Beaked Whale

Mesoplodon grayi
Also known as scamperdown beaked whale, southern beaked whale, baleine à bec de Gray, mésoplodon de Gray, zifio de Gray, ballena de pico de Gray

Length
Males: Up to 18 ft 5 in (5.6 m)
Females: Up to 17 ft 5 in (5.3 m)
Calves at birth: 6 ft 11 in–
7 ft 3 in (2.1–2.2 m)

Adult weight: At least 3,609 lb
(1,100 kg)

Habitat and range: Deep offshore
waters of the cooler temperate
Southern Hemisphere.

Diet: Cephalopods.

Social notes: Mostly seen alone
or in pairs; some mass strandings
are known, indicating that larger
groups may occur.

Conservation status: The IUCN
Red List status is Data Deficient.

Hector's Beaked Whale

Mesoplodon hectori
Also known as skew-beaked whale, *mésoplodon de Hector*, *baleine à bec de Hector*, *zifio de Héctor*, *ballena picuda de Héctor*, *ballena de pico de Héctor*

Length
Males: Up to 14 ft 1 in (4.3 m)
Females: Up to 14 ft 1 in (4.3 m)
Calves at birth: 6 ft 3 in–6 ft 7 in
(1.9–2.0 m)

Adult weight: Unknown

Habitat and range: Deep, cooler
offshore waters of the temperate
Southern Hemisphere.

Diet: Squid.

Social notes: Unknown, only one
confirmed sighting at sea.

Conservation status: The IUCN
Red List status is Data Deficient.

Deraniyagala's Beaked Whale

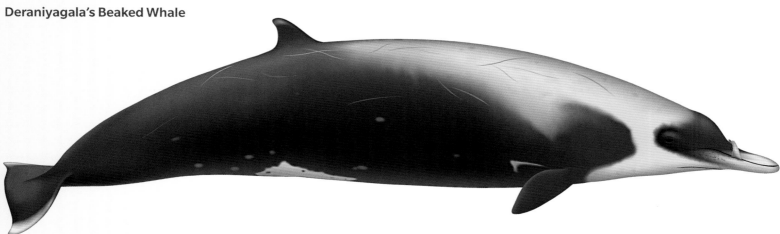

Strap-toothed Beaked Whale

Deraniyagala's Beaked Whale

Mesoplodon hotaula
Also known as *baleine à bec de Deraniyagala, zifio de Deraniyagala*

Length
Males: Up to 15 ft 9 in (4.8 m)
Females: Up to 15 ft 9 in (4.8 m)
Calves at birth (estimated):
6 ft 7 in (2 m)

Adult weight: Unknown

Habitat and range: Deep offshore
waters of the tropical Pacific and
Indian oceans (based on only a
handful of strandings).

Diet: Probably squid.

Social notes: Unknown, never
identified at sea.

Conservation status: The IUCN
Red List status is Data Deficient.

Strap-toothed Beaked Whale

Mesoplodon layardii
Also known as Layard's beaked whale, *baleine à bec de Layard, zifio de
Layard, ballena de pico de Layard*

Length
Males: Up to 20 ft (6.1 m)
Females: Up to 20 ft 4 in (6.2 m)
Calves at birth: 9 ft 10 in (3 m)

Adult weight: Over 2,866 lb
(1,300 kg)

Habitat and range: Deep, cooler
offshore waters of the temperate
Southern Hemisphere.

Diet: Squid, especially small squid
for males.

Social notes: Groups of two to
three are seen.

Conservation status: The IUCN
Red List status is Data Deficient.

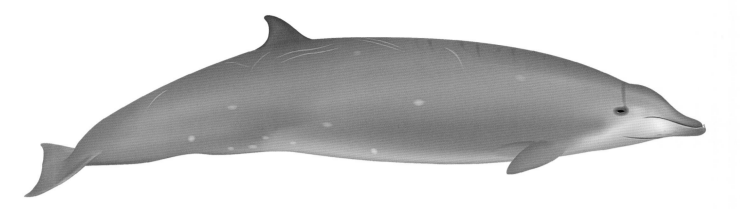

True's Beaked Whale

Perrin's Beaked Whale

True's Beaked Whale

Mesoplodon mirus
Also known as *baleine à bec de True, mésoplodon de True, zifio de True, ballena de pico de True*

Length
Males: Up to 17 ft 9 in (5.4 m)
Females: Up to 17 ft 9 in (5.4 m)
Calves at birth: 6 ft 7 in–8 ft 2 in
(2.0–2.5 m)

Adult weight: Up to 3,086 lb
(1,400 kg)

Habitat and range: Deep offshore waters of the temperate North Atlantic and at least parts of the temperate Southern Hemisphere, based on strandings (disjunct distribution).

Diet: Squid, some fish.

Social notes: Groups of up to three individuals based on a few rare sightings.

Conservation status: The IUCN Red List status is Data Deficient.

Perrin's Beaked Whale

Mesoplodon perrini
Also known as *baleine à bec de Perrin, mésoplodon de Perrin, zifio de Perrin*

Length
Males: Up to 12 ft 10 in (3.9 m)
Females: Up to 14 ft 5 in (4.4 m)
Calves at birth: Unknown

Adult weight: Unknown

Habitat and range: Deep offshore waters of the eastern North Pacific from northern California to Baja California, inclusive.

Diet: Squid.

Social notes: Unknown. Rarely identified at sea, with only a handful of strandings.

Conservation status: The IUCN Red List status is Data Deficient.

Pygmy Beaked Whale

Stejneger's Beaked Whale

Pygmy Beaked Whale

Mesoplodon peruvianus
Also known as Peruvian beaked whale, lesser beaked whale, *baleine à bec pygmée, mésoplodon pygmée, ballena picuda*

Length
Males (estimated): Up to 12 ft 10 in (3.9 m)
Females (estimated): Up to 12 ft 10 in (3.9 m)
Calves at birth (estimated): 5 ft 3 in (1.6 m)

Adult weight: Unknown

Habitat and range: Deep, generally offshore waters of the Eastern Tropical and warm temperate Pacific.

Diet: Squid, midwater fishes, shrimps.

Social notes: Groups of two to five seen, sometimes alone.

Conservation status: The IUCN Red List status is Data Deficient.

Stejneger's Beaked Whale

Mesoplodon stejnegeri
Also known as Bering Sea beaked whale, saber-toothed whale, *mésoplodon de Stejneger, baleine à bec de Stejneger, zifio de Stejneger, ballena de pico de Stejneger*

Length
Males: Up to 18 ft 8 in (5.7 m)
Females: Up to 18 ft 8 in (5.7 m)
Calves at birth: 7 ft 3 in (2.2 m)

Adult weight: At least 3,527 lb (1,600 kg)

Habitat and range: Deep offshore waters of the temperate to subarctic North Pacific.

Diet: Squid.

Social notes: Groups of five and up to 15 individuals.

Conservation status: The IUCN Red List status is Data Deficient.

Spade-toothed Beaked Whale (male)/artist's impression

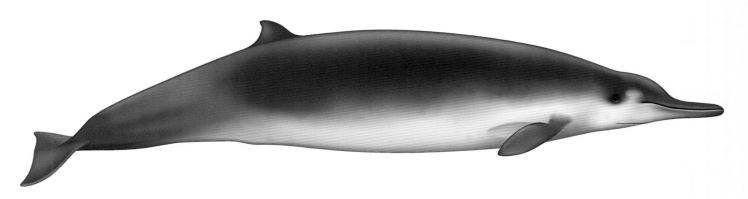

Spade-toothed Beaked Whale (female)

Spade-toothed Beaked Whale

Mesoplodon traversii
Also known as Bahamonde's beaked whale, Traver's beaked whale, *baleine à bec de Travers, mésoplodon de Travers, mésoplodon zifio de Travers, mésoplodon de Bahamonde, zifio de Bahamonde*

Length
Males: Unknown
Females: Up to 17 ft 5 in (5.3 m)
Calves at birth: Unknown.

Adult weight: Unknown

Habitat and range: Deep offshore waters of the subtropical South Pacific and possibly Indian and South Atlantic oceans.

Diet: Unknown but probably squid.

Social notes: Never identified at sea, only known from two stranded specimens.

Conservation status: The IUCN Red List status is Data Deficient.

Blainville's Beaked Whale (male)

Blainville's Beaked Whale (female)

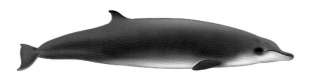

Blainville's Beaked Whale (calf)

Blainville's Beaked Whale

Mesoplodon densirostris
Also known as dense-beaked whale, *mésoplodon de Blainville, baleine à bec de Blainville, zifio de Blainville, ballena picuda de Blainville*

Length
Up to 15 ft 5 in (4.7 m)
Females: Up to 15 ft 5 in (4.7 m)
Calves at birth: 6 ft 7 in–8 ft 2 in
(2.0–2.5 m)

Adult weight: Up to 2,277 lb
(1,033 kg)

Habitat and range: Deep offshore
waters of the temperate and tropical
world ocean.

Social notes: Squid and deep-water
fishes.

Social notes: Usual sightings are of
one or two individuals, but groups
of three to seven are seen; groups
consist of subadults or single males
with multiple females (harems).

Conservation status: The IUCN Red
List status is Data Deficient.

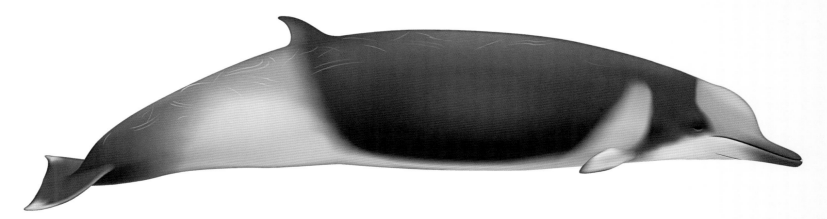

Shepherd's Beaked Whale

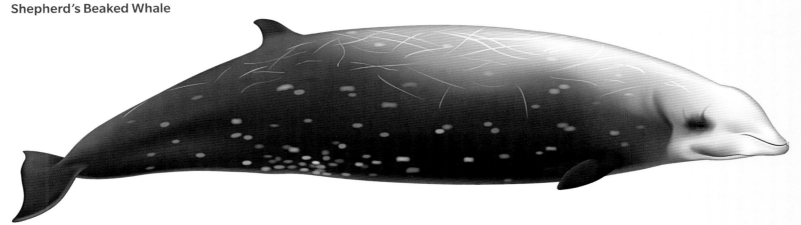

Cuvier's Beaked Whale

Shepherd's Beaked Whale

Tasmacetus shepherdi
Also known as Tasman beaked whale, *tasmacète, ballena picuda de Shepherd*

Length
Males: Up to 23 ft (7 m)
Females: Up to 21 ft 8 in (6.6 m)
Calves at birth (estimated):
9 ft 10 in (3 m)

Adult weight: Unknown

Habitat and range: Cold, deep offshore waters in the temperate Southern Hemisphere (based on limited strandings).

Diet: Fish, squid and crabs.

Social notes: Groups of up to 12 individuals.

Conservation status: The IUCN Red List status is Data Deficient.

Cuvier's Beaked Whale

Ziphius cavirostris
Also known as goose-beaked whale, goosebeak whale, *baleine à bec de Cuvier, baleine de Cuvier, baleine à bec de Cuvier, ziphius, zifio de Cuvier, ballena picuda de Cuvier, ballena de Cuvier*

Length
Males: Up to 23 ft (7 m)
Females: Up to 23 ft (7 m)
Calves at birth: 8 ft 11 in (2.7 m)

Adult weight: Up to 6,614 lb (3,000 kg)

Habitat and range: Deep offshore world ocean, except in Arctic and high latitudes of the Antarctic.

Diet: Mainly squid; also fish and crustaceans.

Social notes: Often alone or in groups of two to seven individuals.

Conservation status: The IUCN Red List status is Least Concern.

Northern Hemisphere Ecotypes

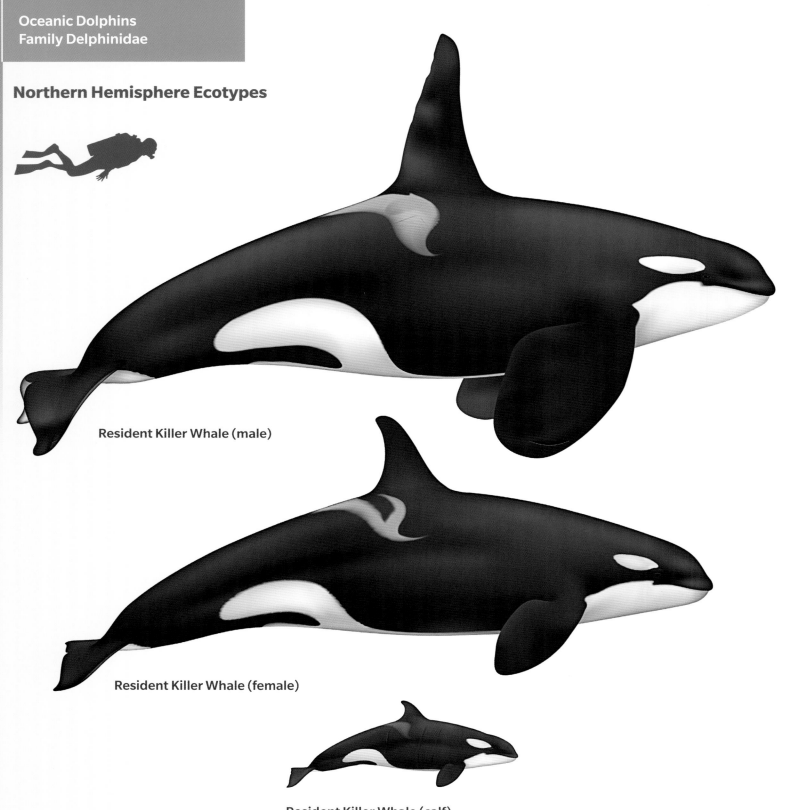

Resident Killer Whale (male)

Resident Killer Whale (female)

Resident Killer Whale (calf)

Northern Hemisphere Ecotypes (continued)

Offshores   **Bigg's Killer Whale (transient)**   **Eastern North Atlantic Type 1**   **Eastern North Atlantic Type 2**

Southern Hemisphere Ecotypes

Antarctic Type A   **Pack Ice Type B**   **Ross Seal Type C**   **Subantarctic Type D**

Killer Whale, or Orca

Orcinus orca
Also known as orca, *orque*, *épaulard*, *espadarte*, *kocatka*

Length
Males: Up to 32 ft 2 in (9.8 m)
Females: Up to 27 ft 11 in (8.5 m)
Calves at birth: 6 ft 11 in–8 ft 6 in
(2.1–2.6 m)

Adult weight: 3.75–5 tons
(7,500–10,000 kg)

Habitat and range: Coastal and offshore world ocean from Antarctic to Arctic, largest numbers found in the Antarctic and in northern North Pacific and North Atlantic waters.

Diet: More than 200 species of fish, cephalopods and marine mammals, including large whales, dolphins, seals, otters, sea lions, penguins; occasionally seabirds.

Social notes: Group size varies by ecotype with marine-mammal-hunting Bigg's (transient) orcas living in smaller groups (in the North Pacific, up to six individuals) and fish-eating orcas found in larger groups of up to 20 individuals.

Larger groupings of up to several hundred sometimes occur. At least eight other ecotypes have been recognized, mainly in the Antarctic.

Conservation status: The IUCN Red List status is Data Deficient. The southern resident fish-eating community of killer whales is an endangered species under U.S. and Canadian laws.

Gerlache Small Type B

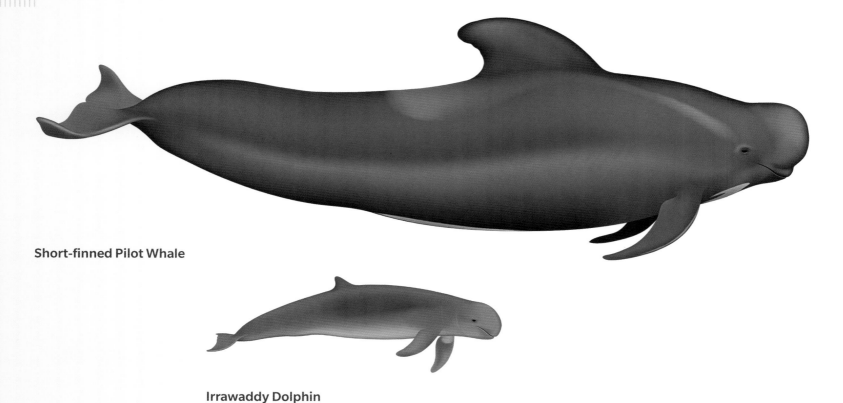

Short-finned Pilot Whale

Irrawaddy Dolphin

Short-finned Pilot Whale

Globicephala macrorhynchus
Also known as pothead, Pacific pilot whale, *globicéphale tropical,*
calderón de aletas cortas, caldrón negro

Length
Males: Up to 23 ft 7 in (7.2 m)
Females: Up to 18 ft (5.5 m)
Calves at birth: 4 ft 7 in–6 ft 3 in
(1.4-1.9 m)

Adult weight: Up to 7,937 lb
(3,600 kg)

Habitat and range: Offshore
tropical and warm temperate world
ocean.

Diet: Mainly squid but also some
fish.

Social notes: Strongly associated
pods of up to several hundred;
travel frequently with other dolphin
species, such as bottlenose,
common and "Lag" dolphins, also
Risso's dolphins and sperm whales.

Conservation status: The IUCN
Red List status is Data Deficient.

Irrawaddy Dolphin

Orcaella brevirostris
Also known as snubfin dolphin, Mahakam River dolphin, *orcelle, delfín*
del Irrawaddy

Length
Males: Up to 9 ft (2.75 m)
Females: Up to 9 ft (2.75 m)
Calves at birth:
3 ft 3 in (1 m)

Adult weight: 254–287 lb
(115–130 kg)

Habitat and range: Coastal
and freshwater tropical eastern
Indian Ocean (Bay of Bengal),
Southeast Asia and the Indonesian
archipelago.

Diet: Various freshwater and
marine fishes, and in marine
populations, squid, cuttlefish and
octopus.

Social notes: Usually six or fewer
but up to 25 in some areas.

Conservation status: The IUCN
Red List status is Vulnerable, but
there are Critically Endangered
populations in the Mekong River,
Malampaya Sound, Mahakam
River, Ayeyarwady River and
Songkhla Lake.

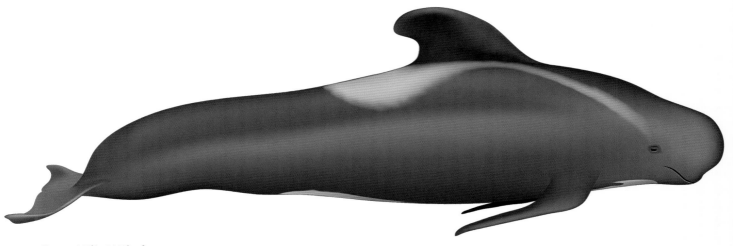

Long-finned Pilot Whale

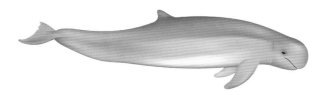

Australian Snubfin Dolphin

Long-finned Pilot Whale

Globicephala melas
Also known as pothead, *globicéphale commun, globicéphale noir, calderón común, dauphin pilote, calderón negro, ballena piloto*

Length
Males: Up to 22 ft (6.7 m)
Females: Up to 18 ft 8 in (5.7 m)
Calves at birth: 5 ft 7 in–5 ft 11 in (1.7–1.8 m)

Adult weight: 2,866–5,071 lb (1,300–2,300 kg)

Habitat and range: Offshore cold temperate North Atlantic and Southern Hemisphere.

Diet: Mainly squid but also fish such as mackerel, herring, hake and cod.

Social notes: Pods of tight, long-lived groups of 20 to a 100 or more, up to more than 1,000.

Conservation status: The IUCN Red List status is Data Deficient.

Australian Snubfin Dolphin

Orcaella heinsohni
Also known as snubfin dolphin, *orcelle d'Australie, delfín del Heinsohn*

Length
Males: Up to 8 ft 10 in (2.7 m)
Females: Up to 8 ft 10 in (2.7 m)
Calves at birth: 3 ft 3 in (1 m)

Adult weight: At least 287 lb (130 kg)

Habitat and range: Tropical northern coastal Australia from Broome in the west to Brisbane, Queensland.

Diet: A wide variety of fishes, as well as squid, cuttlefish, octopus, shrimp and other crustaceans.

Social notes: Group size up to 10, sometimes 20 dolphins.

Conservation status: The IUCN Red List status is Near Threatened.

False Killer Whale

Pygmy Killer Whale

False Killer Whale

Pseudorca crassidens
Also known as pseudorca, *faux-orque, orca falsa*

Length
Males: Up to 19 ft 8 in (6 m)
Females: Up to 16 ft 5 in (5 m)
Calves at birth: 4 ft 11 in–6 ft 11 in
(1.5–2.1 m)

Adult weight: Up to 4,409 lb
(2,000 kg)

Habitat and range: Mainly offshore
in the temperate and tropical
Atlantic, Pacific and Indian oceans.

Diet: Fish, including larger billfish
and mahi-mahi species and various
cephalopods; known to attack
dolphins and sperm whales.

Social notes: Pods of up to 60
and sometimes hundreds closely
associated.

Conservation status: The IUCN
Red List status is Data Deficient.

Pygmy Killer Whale

Feresa attenuata
**Also known as slender blackfish, *épaulard pygmée, orque pygmée, orca
pigmea***

Length
Males: Up to 8 ft 6 in (2.6 m)
Females: Up to 8 ft 6 in (2.6 m)
Calves at birth: 2 ft 8 in (0.8 m)

Adult weight: Up to 496 lb (225 kg)

Habitat and range: Offshore
tropical and subtropical Atlantic,
Pacific and Indian oceans.

Diet: Fish and squid; occasionally
attacks other dolphins.

Social notes: Pods of 12 to 50 and
up to several hundred travel and
hunt together.

Conservation status: The IUCN
Red List status is Data Deficient.

Melon-headed Whale

Risso's Dolphin

Melon-headed Whale

Peponocephala electra
Also known as melon-head, *péponocéphale, calderón pequeño, delfín cabeza de melon*

Length
Males: Up to 9 ft 2½ in (2.78 m)
Females: Up to 9 ft 2½ in (2.78 m)
Calves at birth: 3 ft 3 in (1 m)

Adult weight: Up to 606 lb (275 kg)

Habitat and range: Offshore tropical and subtropical Atlantic, Pacific and Indian oceans.

Diet: Squid, small fishes, occasionally shrimp.

Social notes: Large pods of 100 to 500 individuals, sometimes up to 2,000 or more.

Conservation status: The IUCN Red List status is Least Concern.

Risso's Dolphin

Grampus griseus
Also known as grey dolphin, grampus, *dauphin de Risso, delfín de Risso, dauphin de Risso, calderón gris*

Length
Males: Up to 12 ft 6 in (3.8 m)
Females: Up to 12 ft 6 in (3.8 m)
Calves at birth:
3 ft 7 in–4 ft 11 in (1.1–1.5 m)

Adult weight: Up to 882 lb (400 kg)

Habitat and range: Mainly offshore on continental slope and deeper waters of temperate and tropical Atlantic, Pacific and Indian oceans.

Diet: Squid, octopus and crustaceans.

Social notes: Groups of 10 to a 100 but occasionally thousands; sociable with other dolphin species in different parts of the world, as well as with sperm whales and sometimes with gray whales in the eastern North Pacific.

Conservation status: The IUCN Red List status is Least Concern.

Guiana Dolphin

Tucuxi

Indo-Pacific Bottlenose Dolphin

Guiana Dolphin

Sotalia guianensis
Also known as *costero, sotalia, tonina, bufeo negro, boto común, golfinho cinza, boto*

Length
Males: Up to 5 ft 7 in (1.7 m)
Females: Up to 6 ft 1½ in (1.87 m)
Calves at birth: 2 ft 8 in–3 ft 9 in (0.8–1.15 m)

Adult weight: At least 267 lb (121 kg)

Habitat and range: Nearshore shallow waters and estuaries of coastal Central and tropical South America.

Diet: A wide variety of small bottom-feeding as well as open-ocean fishes; also cephalopods, shrimps and crabs.

Social notes: Groups of one to 40 individuals. In southeastern Brazil, most common grouping was two to 10 individuals. They live in fission-fusion societies with a promiscuous mating system.

Conservation status: The IUCN Red List status is Data Deficient.

Tucuxi

Sotalia fluviatilis
Also known as Guianian river dolphin, estuarine dolphin, gray dolphin, *sotalia, sotalie, dauphin de L'Amazon, bufeo gris, bufeo negro, bufo negro*

Length
Males: Up to 4 ft 11 in (1.49 m)
Females: Up to 5 ft (1.52 m)
Calves at birth: 2 ft 4 in–2 ft 8 in (0.7–0.8 m)

Adult weight: Unknown

Habitat and range: Freshwater distribution throughout the Amazon basin and part of the Orinoco.

Diet: A wide variety of freshwater schooling fishes.

Social notes: Groups of two to four individuals and up to 30.

Conservation status: The IUCN Red List status is Data Deficient.

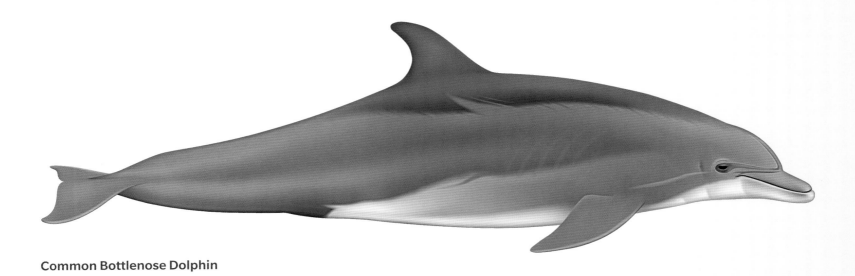

Common Bottlenose Dolphin

Common Bottlenose Dolphin

Tursiops truncatus
Also known as bottlenose dolphin, *grand dauphin, dauphin souffleur, delfín mular, tursión, tonina*

Length
Males: Up to 12 ft 6 in (3.8 m)
Females: Up to 12 ft 6 in (3.8 m)
Calves at birth: 3 ft 3 in–4 ft 3 in (1.0–1.3 m)

Adult weight: Up to 1,433 lb (650 kg)

Habitat and range: Coastal waters, including semi-enclosed seas from nearshore to continental shelf of temperate and tropical Atlantic, Pacific and Indian oceans.

Diet: An extremely wide variety of fishes and squid and sometimes shrimps and other crustaceans.

Social notes: Groups contain fewer than 20 to several hundred individuals, the larger groups mainly found offshore. These are fission-fusion societies with separate young male groups. Separate populations are found inshore and offshore; they often travel with other dolphin species.

Conservation status: The IUCN Red List status is Least Concern, but bottlenose dolphins in the Mediterranean and the Black Sea are rated Endangered and a Fiordland, New Zealand, population is considered Critically Endangered.

Indo-Pacific Bottlenose Dolphin

Tursiops aduncus
Also known as bottlenose dolphin, Indian Ocean bottlenose dolphin, *grand dauphin de l'océan Indien, delfín mular del Océano Indico*

Length
Males: Up to 8 ft 10 in (2.7 m)
Females: Up to 8 ft 10 in (2.7 m)
Calves at birth: 2 ft 10 in–3 ft 8 in (0.85–1.12 m)

Adult weight: Up to 507 lb (230 kg)

Habitat and range: Inshore and coastal waters on the continental shelf of the warm temperate to tropical Indian and western Pacific oceans.

Diet: A wide variety of schooling bottom-feeding and reef fishes, as well as cephalopods.

Social notes: Groups of up to 20, sometimes hundreds; fission-fusion society with separate young male groups.

Conservation status: The IUCN Red List status is Data Deficient.

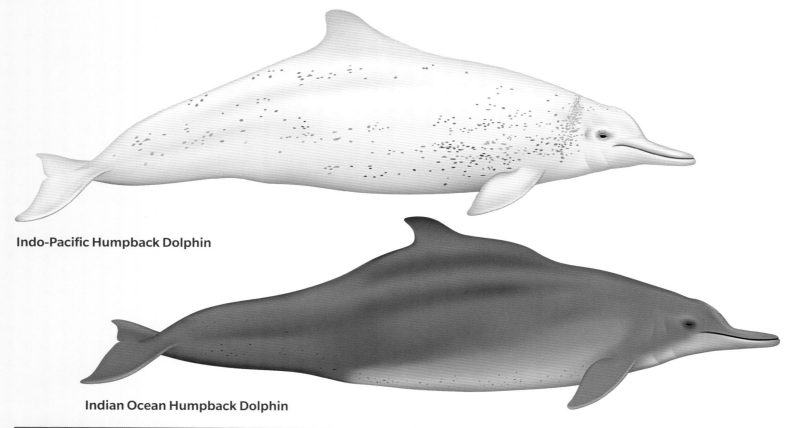

Indo-Pacific Humpback Dolphin

Indian Ocean Humpback Dolphin

Indo-Pacific Humpback Dolphin

Sousa chinensis
Also known as humpbacked dolphin, *dauphin à bosse de l'Indo-Pacifique, delfín jorobado del Indo-Pacífico*

Length
Males: Up to 8 ft 10 in (2.7 m)
Females: Up to 8 ft 6 in (2.6 m)
Calves at birth (estimated):
3 ft 3 in (1 m)

Adult weight: Up to 529 lb (240 kg)

Habitat and range: Nearshore waters (including mangroves and estuaries) of eastern Indian Ocean (Bay of Bengal). Also the waters off southeast Asia and China and the western Indonesian archipelago.

Diet: A variety of nearshore, estuarine and reef fishes and cephalopods.

Social notes: Groups of fewer than 10 and up to 40 individuals.

Conservation status: The IUCN Red List status is Near Threatened and decreasing. The eastern Taiwan Strait population is Critically Endangered.

Indian Ocean Humpback Dolphin

Sousa plumbea
Also known as humpbacked dolphin, *dauphin à bosse de l'océan Indien, delfín jorobado del Océano Índico*

Length
Males: Up to 9 ft 2 in (2.8 m)
Females: Up to 8 ft 6 in (2.6 m)
Calves at birth: 3 ft 3 in (1 m)

Adult weight: Up to 617 lb (280 kg)

Habitat and range: Nearshore tropical and subtropical waters (including mangroves and estuaries) of western and central Indian Ocean from Cape Town, South Africa, to Myanmar on the Bay of Bengal.

Diet: A wide variety of nearshore, estuary and reef fishes and possibly cephalopods.

Social notes: Groups of fewer than 10 and up to more than 100 individuals.

Conservation status: The IUCN Red List status has not been evaluated yet.

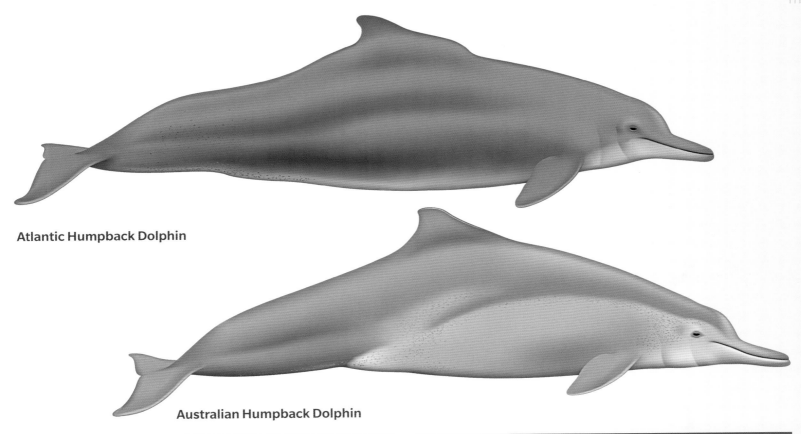

Atlantic Humpback Dolphin

Australian Humpback Dolphin

Atlantic Humpback Dolphin

Sousa teuszii
Also known as humpbacked dolphin, Atlantic hump-backed dolphin, Teusz's dolphin, *dauphin à bosse de l'Atlantique, dauphin du Cameroun, delfín jorobado del Atlántico, delfín blanco Africano, bufeo Africano*

Length
Males: Up to 9 ft 2 in (2.8 m)
Females: Up to 9 ft 2 in (2.8 m)
Calves at birth (estimated):
3 ft 3 in (1 m)

Adult weight: Up to 626 lb (284 kg)

Habitat and range: Nearshore waters of the eastern tropical and subtropical Atlantic Ocean from Angola to Western Sahara.

Diet: Schooling fishes, especially mullet; cooperative herding of fish noted, including in cooperation with human fishers.

Social notes: Groups of two to 10, up to 40 individuals; they sometimes travel and associate with bottlenose dolphins.

Conservation status: The IUCN Red List status is Vulnerable and decreasing.

Australian Humpback Dolphin

Sousa sahulensis
Also known as humpbacked dolphin, *dauphin à bosse d'Australie, delfín jorobado Australiano*

Length
Males: Up to 8 ft 10 in (2.7 m)
Females: Up to 8 ft 10 in (2.7 m)
Calves at birth: 3 ft 3 in (1 m)

Adult weight: At least 529 lb (240 kg)

Habitat and range: Nearshore tropical and subtropical waters (including mangroves and estuaries) of Australia from northern New South Wales to Ningaloo Reef, Western Australia, with populations on the southeast and northwest ends of the island of New Guinea.

Diet: A wide variety of nearshore, estuary and reef fishes and possibly cephalopods.

Social notes: Groups of fewer than 10 individuals and an indication of fluidity in the social structure.

Conservation status: The IUCN Red List status has not been evaluated yet.

Rough-toothed Dolphin

Pantropical Spotted Dolphin

Rough-toothed Dolphin

Steno bredanensis
Also known as *sténo, esteno, delfín de dientes rugosos, delfín de pico largo*

Length
Males: Up to 9 ft 2 in (2.8 m)
Females: Up to 8 ft 8 in (2.65 m)
Calves at birth (estimated):
3 ft 3 in (1 m)

Adult weight: Up to 342 lb (155 kg)

Habitat and range: Deep offshore waters of the tropical and subtropical Atlantic, Pacific and Indian oceans.

Diet: A variety of fishes, especially mahi-mahi; also cephalopods.

Social notes: Groups of 10 to 20 individuals and occasional aggregations of more than 100. These dolphins sometimes travel with pilot whales and bottlenose or other dolphins.

Conservation status: The IUCN Red List status is Least Concern.

Pantropical Spotted Dolphin

Stenella attenuata
Also known as spotter, bridled dolphin, narrow-snouted dolphin, *dauphin tacheté pantropical, estenela moteada, delfín manchado pantropical, delfín pintado*

Length
Males: Up to 8 ft 6 in (2.6 m)
Females: Up to 7 ft 11 in
(2.4 m)
Calves at birth: 2 ft 8 in–2 ft 10 in
(0.8–0.85 m)

Adult weight: Up to 262 lb (119 kg)

Habitat and range: Coastal and offshore waters of the tropical and subtropical Atlantic, Pacific and Indian oceans.

Diet: Various surface to midwater fishes, depending on location; also squids and crustaceans.

Social notes: Coastal schools contain fewer than 100 individuals; in offshore areas, the schools number in the thousands.

Conservation status: The IUCN Red List status is Least Concern. Pantropical spotted dolphins have several geographic forms that are considered subspecies. Some have more spots than others; a few are hardly spotted at all.

Atlantic Spotted Dolphin

Spinner Dolphin

Atlantic Spotted Dolphin

Stenella frontalis
Also known as spotted dolphin, Atlantic spotter, bridled dolphin,
dauphin tacheté de l'Atlantique, delfín pintado, delfín machado del
Atlántico

Length
Males: Up to 7 ft 7 in (2.3 m)
Females: Up to 7 ft 7 in (2.3 m)
Calves at birth:
2 ft 8 in–3 ft 11 in (0.8–1.2 m)

Adult weight: Up to 315 lb (143 kg)

Habitat and range: Offshore
waters of the tropical and warm
temperate Atlantic Ocean.

Diet: A wide variety of surface and
middle layer fishes, squids and
benthic invertebrates.

Social notes: Groups of five to
15 and up to 50 individuals; fluid
group structure.

Conservation status: The IUCN
Red List status is Data Deficient.

Spinner Dolphin

Stenella longirostris
Also known as spinner, long-beaked dolphin, long-snouted dolphin,
dauphin longirostre, estenela giradora, delfín girador, delfín tornillón

Length
Males: Up to 7 ft 9 in (2.35 m)
Females: Up to 6 ft 7 in (2 m)
Calves at birth:
2 ft 6 in–2 ft 8 in (0.75–0.8 m)

Adult weight: Up to 181 lb (82 kg)

Habitat and range: Shallow
coastal waters (resting areas) to
offshore waters of the tropical and
subtropical Atlantic, Pacific and
Indian oceans.

Diet: Wide variety of midwater
fishes, shrimps, squids.

Social notes: Schools range in size
from fewer than 50 to thousands.

Conservation status: The IUCN
Red List status is Data Deficient.

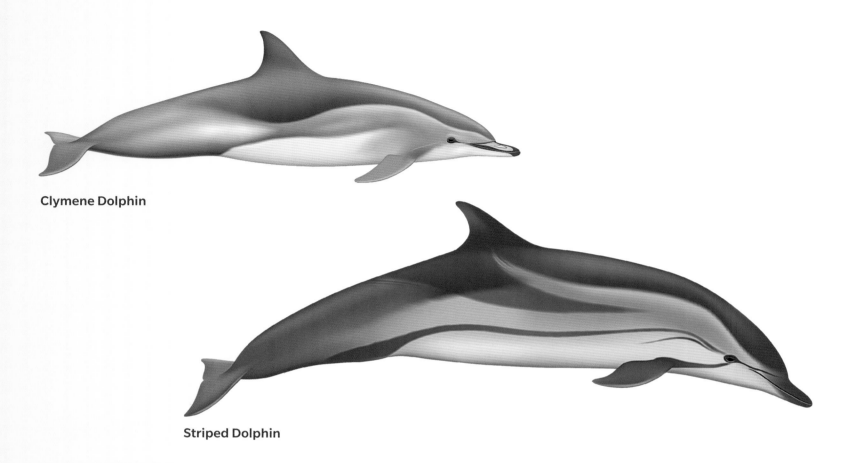

Clymene Dolphin

Striped Dolphin

Clymene Dolphin

Stenella clymene
Also known as short-snouted spinner dolphin, helmet dolphin, Atlantic spinner dolphin, *dauphin de Clymène, delfín Clymene*

Length
Males: Up to 6 ft 6 in (1.97 m)
Females: Up to 6 ft 3 in (1.9 m)
Calves at birth: Up to 3 ft 11 in (1.2 m)

Adult weight: At least 176 lb (80 kg)

Habitat and range: Deep, offshore waters of the tropical and subtropical Atlantic Ocean.

Diet: Small fish and squid.

Social notes: Groups of 60 or more individuals, up to 200, with schools segregated by age and sex; associate with common and spinner dolphins.

Conservation status: The IUCN Red List status is Data Deficient.

Striped Dolphin

Stenella coeruleoalba
Also known as Euphrosyne dolphin, *dauphin bleu et blanc, dauphin rayé, estenela listada, delfín listado, delfín blanco y azul*

Length
Males: Up to 8 ft 5 in (2.56 m)
Females: Up to 8 ft 5 in (2.56 m)
Calves at birth: 3 ft 1 in–3 ft 3 in (0.93–1.0 m)

Adult weight: Up to 344 lb (156 kg)

Habitat and range: Deep offshore waters of the tropical to temperate Atlantic, Pacific and Indian oceans.

Diet: A wide variety of small bottom and midwater fishes such as lanternfish, cod and squids.

Social notes: Groups of 30 to 500 individuals up to several thousand.

Conservation status: The IUCN Red List status is Least Concern

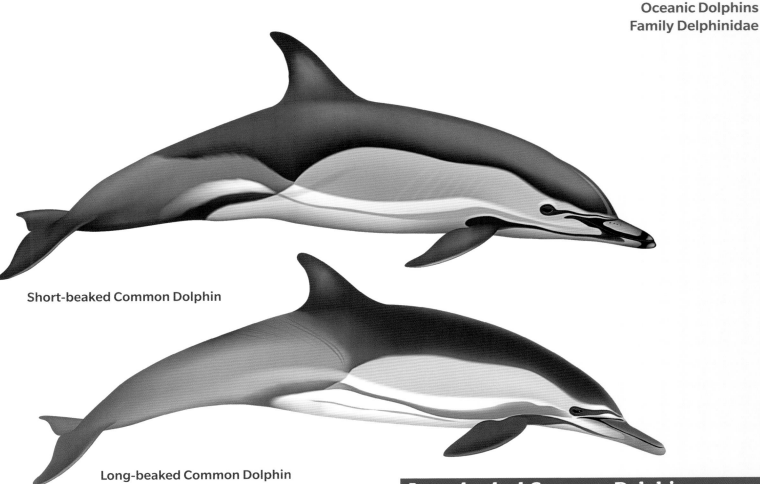

Short-beaked Common Dolphin

Long-beaked Common Dolphin

Short-beaked Common Dolphin

Delphinus delphis
Also known as common, short-beaked saddleback dolphin, saddle-backed dolphin, Atlantic dolphin, Pacific dolphin, *dauphin commun*, *delfín común*, *delfín común de rostro corto*

Length
Males: Up to 8 ft 10 in (2.7 m)
Females: Up to 8 ft 10 in (2.7 m)
Calves at birth: 2 ft 8 in–3 ft
(0.8–0.9 m)

Adult weight: Up to 441 lb (200 kg)

Habitat and range: Nearshore to offshore waters of the temperate to tropical Atlantic and Pacific oceans.

Diet: Small schooling fishes, squid.

Social notes: Schools can be small, fewer than 10, in certain areas, e.g., the Mediterranean, while in the Pacific more than 10,000 sometimes travel together.

Conservation status: The IUCN Red List status is Least Concern globally, but the status of the Mediterranean population is Endangered and the Black Sea population is Vulnerable.

Long-beaked Common Dolphin

Delphinus capensis
Also known as common, *dauphin commun à long bec*, *delfín común de rostro largo*, *delfín común a pico largo*

Length
Males: Up to 8 ft 6 in (2.6 m)
Females: Up to 7 ft 3 in (2.22 m)
Calves at birth (estimated):
2 ft 8 in–3 ft (0.8–0.9 m)

Adult weight: At least 518 lb (235 kg)

Habitat and range: Nearshore and generally coastal waters of the tropical Atlantic, Pacific and Indian oceans. Many gaps in distribution may be partly due to confusing identification with the closely related short-beaked common dolphin.

Diet: A wide variety of small schooling fishes such as sardines, anchovies and pilchards; squids.

Social notes: Groups of 10 to several thousand, with some age and sex segregation.

Conservation status: The IUCN Red List status is Data Deficient.

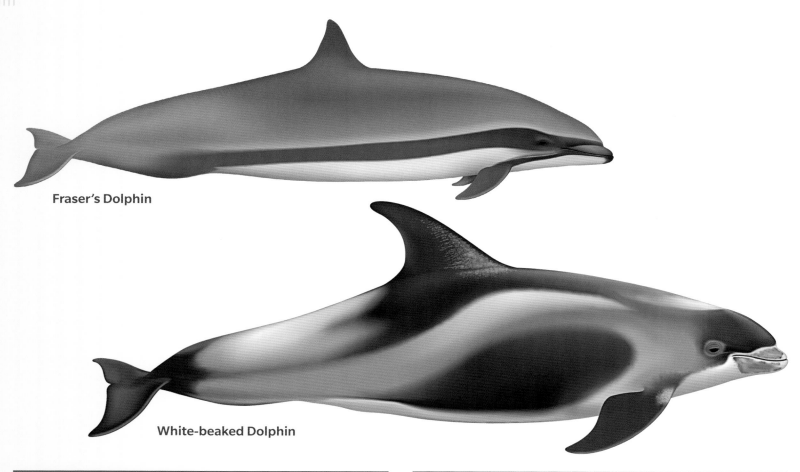

Fraser's Dolphin

White-beaked Dolphin

Fraser's Dolphin

Lagenodelphis hosei
Also known as Sarawak dolphin, *dauphin de Fraser, delfín de Fraser, delfín de Borneo*

Length
Males: Up to 8 ft 10 in (2.7 m)
Females: Up to 8 ft 6 in (2.6 m)
Calves at birth: 3 ft 3 in–3 ft 7 in
(1.0–1.1 m)

Adult weight: At least 463 lb
(210 kg)

Habitat and range: Deep offshore
waters of the warm temperate and
tropical Atlantic, Pacific and Indian
oceans; nearshore in areas with
deep waters close to land.

Diet: Midwater fishes, especially
lanternfish, squid and crustaceans.

Social notes: Groups of a hundred
to thousands; often associated with
other dolphin species.

Conservation status: The IUCN
Red List status is Least Concern.

White-beaked Dolphin

Lagenorhynchus albirostris
Also known as *dauphin à bec blanc, lagénorhynque à bec blanc de
l'Atlantique, delfín de hocico blanco, delfín de pico blanco*

Length
Males: Up to 10 ft 2 in (3.1 m)
Females: Up to 10 ft 2 in (3.1 m)
Calves at birth: 3 ft 7 in–3 ft 11 in
(1.1–1.2 m)

Adult weight: 397–772 lb
(180–350 kg)

Habitat and range: Coastal to
offshore deep waters of the cold
temperate to subarctic North
Atlantic.

Diet: A variety of small schooling
fishes, such as herring, cod,
haddock, hake and whiting, as well
as squids and crustaceans.

Social notes: Groups of up to 30
dolphins are typical, but sometimes
hundreds or thousands join
together. There is some evidence of
sex and age segregation.

Conservation status: The IUCN
Red List status is Least Concern.

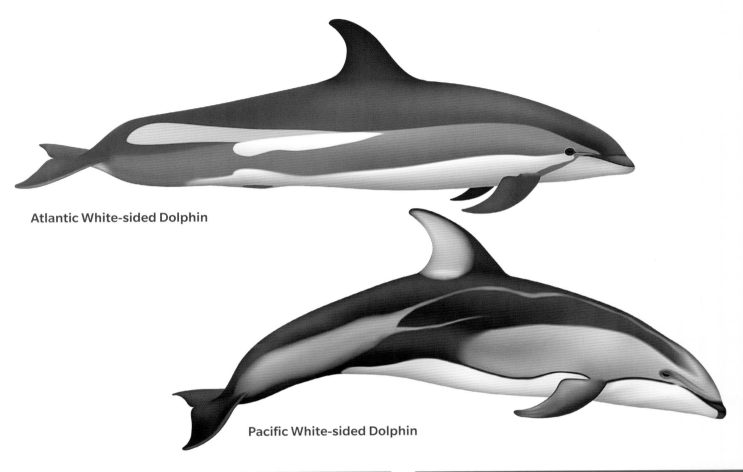

Atlantic White-sided Dolphin

Pacific White-sided Dolphin

Atlantic White-sided Dolphin

Lagenorhynchus acutus
Also known as white-sided dolphin, *dauphin à flancs blancs de l'Atlantique, lagénorhynque à flanc blanc de l'Atlantique, delfín de flancos blancos, delfín de costados blancos*

Length
Males: Up to 9 ft 2 in (2.8 m)
Females: Up to 8 ft 2 in (2.5 m)
Calves at birth: 3 ft 7 in–3 ft 11 in (1.1–1.2 m)

Adult weight: 401–518 lb (182–235 kg)

Habitat and range: Coastal inshore to deep offshore waters of the cold temperate and subarctic North Atlantic.

Diet: Small schooling fishes such as herring, mackerel and sand lance; also shrimp and squid.

Social notes: Average group size of 50, but hundreds and sometimes thousands travel together; they sometimes feed with other dolphins and humpback whales.

Conservation status: The IUCN Red List status is Least Concern.

Pacific White-sided Dolphin

Lagenorhynchus obliquidens
Also known as white-sided dolphin, *dauphin à flancs blancs du Pacifique, lagénorhynque à flanc blanc du Pacifique, delfín de costados blancos del Pacífico*

Length
Males: Up to 8 ft 2 in (2.5 m)
Females: Up to 7 ft 11 in (2.4 m)
Calves at birth:
3 ft 0.2 in–3 ft 3 in (0.92–1.0 m)

Adult weight: Up to 437 lb (198 kg)

Habitat and range: Nearshore to deep offshore waters of the cool temperate North Pacific.

Diet: Small schooling fishes, such as lanternfish, anchovies and hake, as well as cephalopods.

Social notes: Groups of 10 and up to hundreds and even thousands, often segregated by age and sex; Risso's and northern right whale dolphins, as well as other dolphins, often join up with the large groups.

Conservation status: The IUCN Red List status is Least Concern.

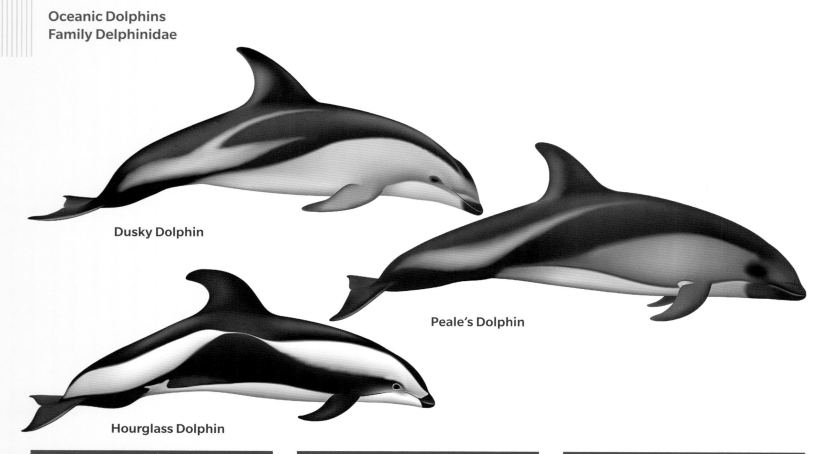

Dusky Dolphin

Peale's Dolphin

Hourglass Dolphin

Dusky Dolphin

Lagenorhynchus obscurus
Also known as dusky, *dauphin somber,*
lagénorhynque somber, delfín oscuro, delfín
listado

Length
Males: Up to 6 ft 11 in (2.1 m)
Females: Up to 6 ft 11 in (2.1 m)
Calves at birth: 2 ft 8 in–3 ft 3 in (0.8–1.0 m)

Adult weight: 154–187 lb (70–85 kg)

Habitat and range: Coastal to continental slope
waters of the Southern Hemisphere, including
some tropical and cold temperate waters.

Diet: Southern anchovy, as well as wide variety
of midwater and benthic fishes; squid.

Social notes: Typical group size is 20 or more,
occasionally 500 to 1,000; feeding may be
coordinated through leaping.

Conservation status: The IUCN Red List status
is Data Deficient.

Hourglass Dolphin

Lagenorhynchus cruciger
Also known as *dauphin crucigère,*
lagénorhynque crucigère, delfín cruzado

Length
Males: Up to 6 ft 3 in (1.9 m)
Females: Up to 5 ft 11 in (1.8 m)
Calves at birth (estimated):
3 ft 3 in (1 m)

Adult weight: 194–207 (88–94 kg)

Habitat and range: Deep offshore, cold
temperate to Antarctic circumpolar waters of
the Southern Hemisphere, extending to the
ice.

Diet: Small fishes, squids and crustaceans.

Social notes: Groups of up to eight; sometimes
assemblages of 60 or more.

Conservation status: The IUCN Red List status
is Least Concern.

Peale's Dolphin

Lagenorhynchus australis
Also known as blackchin dolphin, *dauphin de*
Peale, lagénorhynque de Peale, delfín austral

Length
Males: Up to 7 ft 3 in (2.2 m)
Females: Up to 6 ft 11 in (2.1 m)
Calves at birth: 3 ft 3 in (1 m)

Adult weight: Up to 254 lb (115 kg)

Habitat and range: Coastal and inshore, as
well as sometimes offshore, cold temperate
waters of southern Chile and Argentina,
including the Drake Passage and the Falkland
Islands (Las Malvinas).

Diet: Bottom-feeding fishes, squid and octopus.

Social notes: Groups of five to 30, sometimes
up to 100.

Conservation status: The IUCN Red List status
is Data Deficient.

Northern Right Whale Dolphin

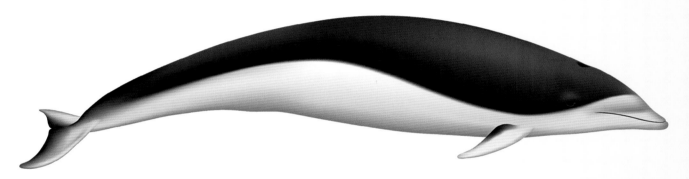

Southern Right Whale Dolphin

Northern Right Whale Dolphin

Lissodelphis borealis
Also known as *dauphin à dos lisse boréal, dauphin aptère boréal, delfín liso del norte*

Length
Males: Up to 10 ft 2 in (3.1 m)
Females: Up to 7 ft 7 in (2.3 m)
Calves at birth: 3 ft 3 in (1 m)

Adult weight: Up to 254 lb (115 kg)

Habitat and range: Deep offshore waters of the temperate North Pacific.

Diet: Squid, lanternfish, hake and various surface and midwater fishes.

Social notes: Groups of 100 to 200 are common but sometimes numbers may go up to 3,000 individuals; often mix with other dolphins.

Conservation status: The IUCN Red List status is Least Concern.

Southern Right Whale Dolphin

Lissodelphis peronii
Also known as *dauphin aptère austral, delfín liso austral, tonina sin aleta*

Length
Males: Up to 9 ft 10 in (3 m)
Females: Up to 9 ft 10 in (3 m)
Calves at birth: 3 ft 3 in (1 m)

Adult weight: Up to 256 lb (116 kg)

Habitat and range: Deep offshore waters of the cool temperate to subantarctic waters of the Southern Hemisphere.

Diet: Fish, including lanternfish, plus squid.

Social notes: Groups of up to more than 1,000 individuals.

Conservation status: The IUCN Red List status is Data Deficient.

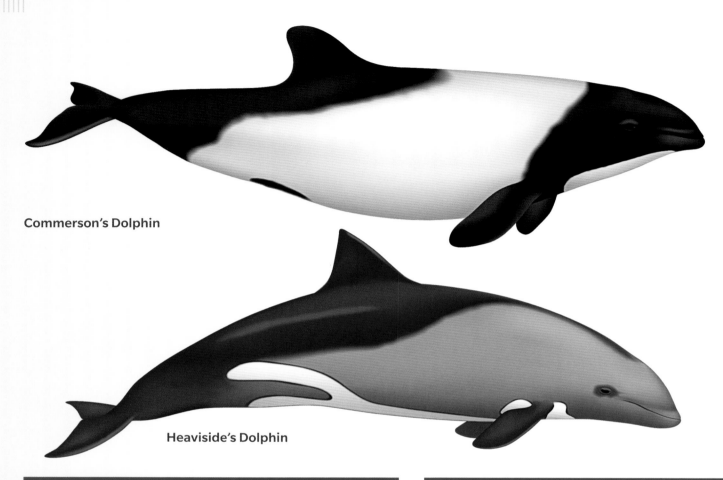

Commerson's Dolphin

Heaviside's Dolphin

Commerson's Dolphin

Cephalorhynchus commersonii
Also known as piebald dolphin, *dauphin de Commerson, tonina overa, tonina dolphin, delfín de Commerson, jacobita*

Length
Males: Up to 5 ft 11 in (1.8 m)
Females: Up to 5 ft 11 in (1.8 m)
Calves at birth:
2 ft 2 in–2 ft 6 in (0.65–0.75 m)

Adult weight: Up to 190 lb (86 kg)

Habitat and range: Coastal waters, including estuaries, of Argentina, southernmost Chile, the offshore Falkland Islands (Las Malvinas) and the French Southern Territory of Kerguelen.

Diet: Opportunistic bottom feeders of a variety of fishes, cephalopods, crustaceans and small invertebrates.

Social notes: In most groups, there are fewer than 10 individuals, but there are sometimes up to more than 100.

Conservation status: The IUCN Red List status is Data Deficient.

Heaviside's Dolphin

Cephalorhynchus heavisidii
Also known as Haviside's dolphin, South African dolphin, Benguela dolphin, *dauphin de Heaviside, céphalorhynque du cap, delfín del Cabo, tonina de Heaviside*

Length
Males: Up to 5 ft 7 in (1.7 m)
Females: Up to 5 ft 7 in (1.7 m)
Calves at birth:
2 ft 8 in–2 ft 10 in (0.8–0.85 m)

Adult weight: 165 lb (75 kg)

Habitat and range: Coastal waters of Namibia, western South Africa, especially west of Cape Town, up to and including the coast of Angola.

Diet: Bottom-dwelling fishes, including hake as well as various schooling fishes and cephalopods.

Social notes: Groups are fewer than 10 individuals and often only two or threes.

Conservation status: The IUCN Red List status given is Data Deficient.

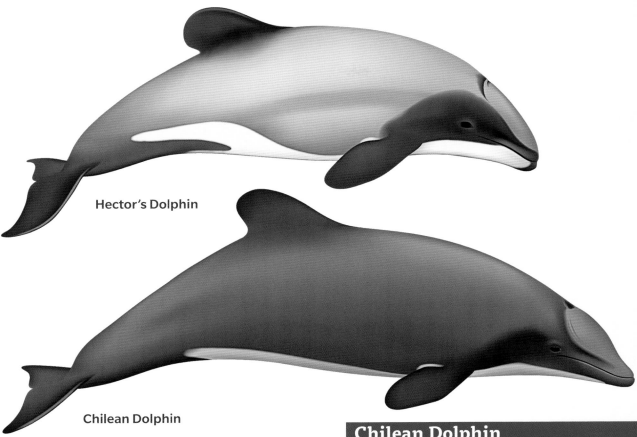

Hector's Dolphin

Chilean Dolphin

Hector's Dolphin

Cephalorhynchus hectori
Also known as New Zealand dolphin, white-headed dolphin, *dauphin d'Hector*, *delfín de Héctor*, *tunina de Héctor*; Maui's (Maui) dolphin (one of four populations)

Length
Males: Up to 4 ft 10 in (1.46 m)
Females: Up to 5 ft 4 in (1.63 m)
Calves at birth: 2 ft–2 ft 4 in
(0.6–0.7 m)

Adult weight: Up to 126 lb (57 kg)

Habitat and range: Nearshore and coastal waters, mainly less than 330 ft (100 m) depth around the coast of New Zealand.

Diet: Various species of small fish and squid.

Social notes: Groups are usually two to eight dolphins but temporary aggregations of 50 or more are sometimes seen.

Conservation status: The IUCN Red List status is Endangered for the species as a whole and Critically Endangered for the Maui's dolphin subspecies. Their present and continuing decline is due to the high numbers of bycatch that result from set net and trawl fishing in nearshore waters.

Chilean Dolphin

Cephalorhynchus eutropia
Also known as black dolphin, black Chilean dolphin, *dauphin du Chili*, *delfín chileno*, *delfín negro*, *tonina de vientre blanco*

Length
Males: Up to 5 ft 7 in (1.7 m)
Females: Up to 5 ft 7 in (1.7 m)
Calves at birth (estimated): 3 ft 3 in
(1 m)

Adult weight: At least 139 lb (63 kg)

Habitat and range: Nearshore and coastal waters of Chile, mainly in waters less than 65 ft (20 m) deep, ranging into Argentine waters off Tierra del Fuego and southern Argentina.

Diet: Shallow-water fishes, including sardines, anchovies, crustaceans and cephalopods.

Social notes: Groups of two to 15 seen in prime areas; sometimes associate in groups larger than 20; may travel in even larger mixed groups with Peale's dolphins.

Conservation status: The IUCN Red List status is Near Threatened based on limited data. The species is probably declining, for Chilean dolphins were hunted for food and crab bait for many years until the 1980s; they are also threatened by gillnet catches and aquaculture farms.

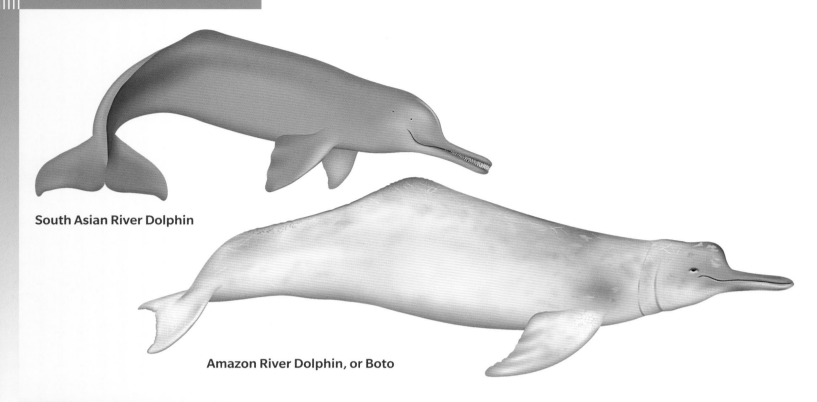

South Asian River Dolphin

Amazon River Dolphin, or Boto

South Asian River Dolphin
Family Platanistidae

Platanista gangetica
Also known as Ganges River dolphin, Ganges dolphin, *susu*, *Ganges susu*, *shushuk*, *plataniste du Gange*, *delfín del Ganges*; Indus River dolphin, *bhulan*, *sousou*

Length
Males: Up to 7 ft 3 in (2.2 m)
Females: Up to 8 ft 6 in (2.6 m)
Calves at birth: 2 ft 4 in–3 ft
(0.7–0.9 m)

Adult weight: At least 187 lb
(85 kg)

Habitat and range: Main river channels, seasonal tributaries and lakes of the Ganges-Brahmaputra-Megna river basin and the Karnaphuli-Sangu river systems and many of their tributaries in India, Bangladesh and Nepal; Pakistan's Indus River.

Diet: A variety of freshwater fishes, mainly bottom fish, and some invertebrates such as prawns.

Social notes: Groups of fewer than 10 individuals but some loose aggregations of up to 30.

Conservation status: The IUCN Red List status is Endangered for this species; the Indus River dolphin and Ganges River dolphin, the two main separate populations (considered subspecies) are also rated Endangered.

Amazon River Dolphin, or Boto
Family Iniidae

Inia geoffrensis
Includes Bolivian river dolphin, Araguaian river dolphin, Orinoco River dolphin. Also known as pink river dolphin, *dauphin de l'Amazone*, *inia*, *bufeo*, *delfín del Amazonas*, *boutu*

Length
Males: Up to 9 ft 2 in (2.8 m)
Females: Up to 7 ft 7 in (2.3 m)
Calves at birth: 2 ft 8 in (0.8 m)

Adult weight: 298–456 lb
(135–207 kg)

Habitat and range: Fresh water in the Amazon and Orinoco river systems, including seasonal tributaries.

Diet: A wide variety of freshwater fishes (43 plus species), mainly bottom fish.

Social notes: Usually single, pairs (mother and calf), up to groups of three; occasionally up to 20 are seen in one area.

Conservation status: The IUCN Red List status is Data Deficient.

Baiji

Franciscana

Baiji
Family Lipotidae

Lipotes vexillifer
Also known as Yangtze River dolphin, whitefin dolphin, white flag dolphin, Chinese lake dolphin, Changjiang dolphin, *dauphin fluviatile de Chine*, *delfín de China*

Length
Males: Up to 7 ft 7 in (2.3 m)
Females: Up to 8 ft 6 in (2.6 m)
Calves at birth: 3 ft (0.9 m)

Adult weight: 346–368 lb
(157–167 kg)

Habitat and range: Freshwater in
the Yangtze River.

Diet: A variety of freshwater fish
species.

Social notes: Two to six was typical
group size, up to 16 individuals.

Conservation status: The IUCN Red
List status is Critically Endangered
but no dolphins have been seen
since before 2006, so this species is
now considered Extinct.

Franciscana
Family Pontoporiidae

Pontoporia blainvillei
Also known as La Plata River dolphin, *dauphin de La Plata*, *delfín de la Plata*, *tonina*

Length
Males: Up to 5 ft 4 in (1.63 m)
Females: Up to 5 ft 10 in (1.77 m)
Calves at birth: 2 ft 5 in (0.73 m)

Adult weight: Up to 117 lb (53 kg)

Habitat and range: Shallow,
coastal waters of temperate
western South Atlantic from Brazil
to Argentina.

Diet: A variety of shallow-water fish
(58 species) as well as cephalopods
and crustaceans.

Social notes: Two to six individuals
form the typical group size, which
can reach up to 15 individuals.

Conservation status: The IUCN
Red List status is Vulnerable. The
Uruguayan and southern Brazilian
population is also separately rated
Vulnerable.

Dall's Porpoise

Harbor Porpoise

Dall's Porpoise

Phocoenoides dalli
Also known as *marsouin de Dall, marsopa de Dall*

Length
Males: Up to 7 ft 11 in (2.4 m)
Females: Up to 7 ft 3 in (2.2 m)
Calves at birth: 3 ft 3 in (1 m)

Adult weight: Up to 441 lb (200 kg)

Habitat and range: Deep coastal
and offshore waters of the
temperate to subarctic North
Pacific.

Diet: Opportunistic hunters, with
diet containing a wide range of
surface and midwater fishes,
especially lanternfish and squid.

Social notes: Groups of up to
12 individuals and occasionally
larger groups of a few hundred or
more.

Conservation status: The IUCN
Red List status is Least Concern.

Harbor Porpoise

Phocoena phocoena
Also known as common porpoise, *marsouin commun, marsopa común*

Length
Males: Up to 5 ft 10 in (1.78 m)
Females: Up to 6 ft 2.5 in (1.89 m)
Calves at birth: 2 ft 4 in–3 ft
(0.7–0.9 m)

Adult weight: 99–154 lb
(45–70 kg)

Habitat and range: Shallow
nearshore to offshore waters
of the temperate to subarctic
North Atlantic and North Pacific,
extending into the Arctic Ocean.

Diet: A variety of fish, especially
small schooling fish such as
mackerel and herring; bottom fish
as well as cephalopods.

Social notes: Groups of fewer than
six and occasionally large loose
groups of up to a few hundred are
seen.

Conservation status: The IUCN
Red List status is Least Concern.
In the Black Sea, the harbor
porpoise population is considered
Endangered; in the Baltic Sea, they
are Critically Endangered.

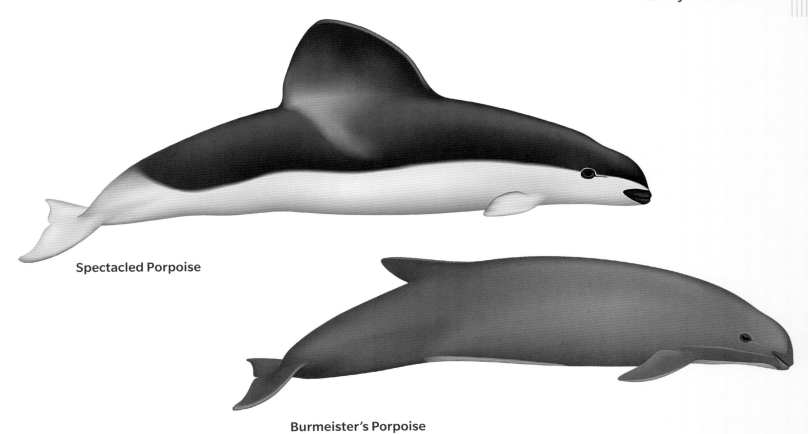

Spectacled Porpoise

Burmeister's Porpoise

Spectacled Porpoise

Phocoena dioptrica
Also known as *marsouin de lahille, marsouin à lunettes, marsopa de anteojos*

Length
Males: Up to 7 ft 4 in (2.24 m)
Females: Up to 6 ft 8 in (2.04 m)
Calves at birth: 3 ft 3 in (1 m)

Adult weight: At least 254 lb (115 kg)

Habitat and range: Coastal and offshore waters of the subantarctic Southern Hemisphere, extending into cold temperate waters of the western South Atlantic as far north as Uruguay.

Diet: Small fish such as anchovies and mantis shrimp.

Social notes: Groups of two to five but individuals often seen traveling alone.

Conservation status: The IUCN Red List status is Data Deficient.

Burmeister's Porpoise

Phocoena spinipinnis
Also known as porpoise, black porpoise, *marsouin de Burmeister, marsopa espinosa*

Length
Males: Up to 6 ft 7 in (2 m)
Females: Up to 6 ft 7 in (2 m)
Calves at birth: 2 ft 8 in–3 ft (0.8–0.9 m)

Adult weight: Up to 187 lb (85 kg)

Habitat and range: Shallow coastal waters of South America on both Pacific (Peru to Chile) and Atlantic (southern Brazil to Tierra del Fuego, Argentina) sides.

Diet: Bottom-feeding as well as pelagic fishes, including anchovies and hake, as well as squid.

Social notes: Groups of up to five or six; sometimes aggregations of up to 70 individuals.

Conservation status: The IUCN Red List status is Data Deficient.

Vaquita

Indo-Pacific Finless Porpoise

Vaquita

Phocoena sinus
Also known as Gulf of California porpoise, Gulf porpoise, *marsouin du golfe de Californie*, cochito

Length
Males: Up to 4 ft 9 in (1.45 m)
Females: Up to 4 ft 11 in (1.5 m)
Calves at birth: 2 ft 4 in (0.7 m)

Adult weight: Unknown

Habitat and range: Shallow waters of northern Gulf of California, especially western portion, limited to core area of 770 sq mi (2,000 sq km).

Diet: A wide variety of bottom fish, including grunts and croakers, plus squids and crustaceans.

Social notes: Small groups, often two but up to eight to 10 individuals.

Conservation status: The IUCN Red List status is Critically Endangered. This is the most endangered cetacean in the world, the one closest to extinction.

Indo-Pacific Finless Porpoise

Neophocaena phocaenoides
Also known as finless porpoise, *marsouin aptère*, marsopa lisa o sin aleta

Length
Males: Up to 5 ft 7 in (1.7 m)
Females: Up to 5 ft 7 in (1.7 m)
Calves at birth: 2 ft 6 in–2 ft 10 in (0.75–0.85 m)

Adult weight: Unknown

Habitat and range: Shallow coastal waters of the tropical and warm temperate northern Indian Ocean and western Pacific bordering western Indonesia (including Borneo), as well as Southeast Asia and China.

Diet: Small fishes, squids and crustaceans such as shrimps.

Social notes: Groups of up to 20 individuals are typical; sometimes up to 50 or more.

Conservation status: The IUCN Red List status is Vulnerable and decreasing.

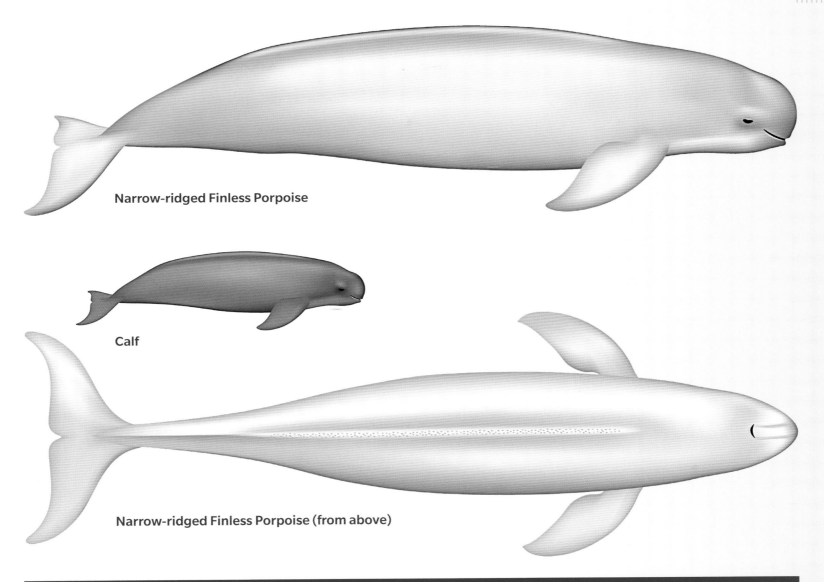

Narrow-ridged Finless Porpoise

Calf

Narrow-ridged Finless Porpoise (from above)

Narrow-ridged Finless Porpoise

Neophocaena asiaorientalis
Also known as finless black porpoise, finless porpoise, *marsouin aptère, marsopa lisa o sin aleta, sunameri*

Length
Males: Up to 7 ft 5 in (2.27 m)
Females: Up to 7 ft 5 in (2.27 m)
Calves at birth: 2 ft 6 in–2 ft 10 in
(0.75–0.85 m)

Adult weight: Unknown

Habitat and range: Shallow coastal waters of the cool temperate western Pacific, including Japan and China; riverine population in the Yangtze River.

Diet: Small fishes, squids and crustaceans such as shrimps.

Social notes: Groups of up to 20 individuals are typical; sometimes up to 50 or more.

Conservation status: The IUCN Red List status is Vulnerable and decreasing.

Getting Involved

Five Ways to Participate in Field Research as a Citizen Scientist

1 Follow Facebook, Twitter and other social-media sites of whale researchers and conservation groups. Facebook sites: Whales.org, Russianorca, American Cetacean Society, Whalewatching Blueprint, Orcatour, The Whale Trail, Hebridean Whale and Dolphin Trust, Irish Whale and Dolphin Group, International Fund for Animal Welfare – IFAW, among many others.

2 Start or join a community-based beach, river or estuary cleanup. Spend time near the water, especially with like-minded people who may be alert for the signs of marine life. Your actions are helping both the human and the cetacean ecosystem.

3 Look for opportunities on MarMam, the list serve for cetacean research and conservation. To subscribe, go to lists.uvic.ca/mailman/listinfo/marmam

4 Join a whale or dolphin research project offered by Earthwatch, United States, U.K. and other countries (earthwatch.org), Tethys Research Institute, Italy (tethys.org), Oceanic Society Expeditions, United States (oceanicsociety.org)

5 To volunteer with a conservation group, consider the groups listed below:
- Whale and Dolphin Conservation (U.K., United States, Argentina, Germany, Australia); Contact: uk.whales.org/support-us/volunteer-with-wdc; info@whales.org
- Cabrillo Whalewatch, San Pedro, CA, United States. Contact: cabrillomarineaquarium.org
- American Cetacean Society, San Francisco, CA, United States. Contact: acsonline.org
- Seymour Marine Discovery Center, Santa Cruz, CA, United States. Contact: seymourcenter.ucsc.edu
- Coastal Research & Education Society of Long Island, Montauk, NY, United States
- Hebridean Whale and Dolphin Trust, Scotland, U.K. Contact: whaledolphintrust.co.uk/get-involved-Volunteering.asp
- Port Townsend Marine Science Center, Port Townsend, WA, United States. Contact: ptmsc.org
- The Whale Trail, a network of sites along the coasts of British Columbia, Canada; and Washington, Oregon and California, United States. Contact: thewhaletrail.org

A gray whale spyhops in the waters off Baja California, Mexico.

Photo Credits

Principal photography
© Brandon Cole

2–3, 6–7, 10, 12–13, 16 (right, top and bottom), 17, 21 (all), 24–25, 26, 29, 32, 36, 40, 47, 50 (all), 51 (both), 56, 57, 63, 65, 66–67, 68, 76–77, 78, 82–83, 86–87, 93, 94 (both), 95, 100–101, 102–103 (all), 104, 108–109, 111, 114–115, 117, 118, 120, 122 (large), 134–135, 138 (left), 139 (right), 143, 144, 148–149, 150–151, 159, 160–161, 163, 164–165 (both), 166, 171, 173, 175, 186, 189, 203, 204–205, 220–221, 222, 295

Other contributors

14 © Tatiana Ivkovich/Far East Russia Orca Project (Whale and Dolphin Conservation)

18 © Mary Evans Picture Library/Alamy Stock Photo

19 © Greenpeace/Rex Weyler

20 © Deron Verbeck/iamaquatic.com

23 © Flip Nicklin/Minden Pictures/Getty Images

28 © Far East Russia Orca Project (Whale and Dolphin Conservation)

30–31 (all) © Russian Cetacean Habitat Project (Whale and Dolphin Conservation) courtesy Olga Titova

35 © Far East Russia Orca Project (Whale and Dolphin Conservation)

39 © Robert Pitman

41 © Far East Russia Orca Project (Whale and Dolphin Conservation)

42–43, 44 © Deron Verbeck/iamaquatic.com

49 (clockwise from top left): © Alisa Schulman-Janiger, © Russian Cetacean Habitat Project (Whale and Dolphin Conservation), © Jean-Pierre Sylvestre, © Charlie Phillips, © Jean-Pierre Sylvestre

53, 54 © Roman Uchytel/Prehistoric Fauna

59 Graphic produced from SPLASH photo ID data, © NOAA: National Marine Sanctuaries

60 © Tim Cuff/Associated Press

70–71 © Mariano Sironi

72 © Can Stock Photo Inc./kenm

74 © North Slope Borough Department of Wildlife Management

81 © Jean-Pierre Sylvestre

85 © Edward Lyman/Hawaiian Islands Humpback Whale National Marine Sanctuary, Large Whale Entanglement Response, from research conducted pursuant to Permit No. 932-1905/MA-009526

88 © Jean-Pierre Sylvestre

91 © Robert L. Pitman/SeaPics.com/archive.org

96, 97, 98 © Jean-Pierre Sylvestre

99 © Rubaiyat Mansur

106 © Robert Pitman

122 (inset), 123 © Jean-Pierre Sylvestre

124 © David Fleetham/Alamy Stock Photo

127 © Russian Cetacean Habitat Project (Whale and Dolphin Conservation)

128 © John D. McHugh/Getty Images

131 © Heiti Paves/123RF

137 © Isuaneye/Dreamstime.com

138 (right) © Elise V/Shutterstock

139 (left) © Alicia Chelini/Shutterstock

145 © Tatiana Ivkovich/Far East Russia Orca Project (Whale and Dolphin Conservation)

146 © Far East Russia Orca Project (Whale and Dolphin Conservation)

152 © Kike Calvo/Alamy Stock Photo

154 © Anthony Pierce/Alamy Stock Photo

156, 158 © Charlie Phillips

168 © Alisa Schulman-Janiger

172 Map © Lesley Frampton, Mike Bossley and Erich Hoyt/Whale and Dolphin Conservation

176 © Jean-Pierre Sylvestre

179 © Tatiana Ivkovich, Far East Russia Orca Project (Whale and Dolphin Conservation)

184–185 © Deron Verbeck/iamaquatic.com

188 © Kyodo/Associated Press

190 © Steve Morgan/Alamy Stock Photo

193 © Jean-Pierre Sylvestre

194–195 © Rob Lott

196 © Robert Pitman

197 © Hubert Yann/Alamy Stock Photo

199 © Angelo Gandolfi/naturepl.com

200 © Alisa Schulman-Janiger

206–207 © Rich Carey/Shutterstock

208 © Wild Horizon/Getty Images

211 © FLPA/Alamy Stock Photo

213 Map © Lesley Frampton, Giuseppe Notarbartolo di Sciara, Mike Tetley and Erich Hoyt/IUCN Marine Mammal Protected Areas Task Force

214, 217 © Mike Bossley

218 © Charlie Phillips

Sources and Resources

Acomprehensive list of references used in the preparation of this volume would cover many pages. Below are the main references used, as well as good sources for further reading.

Berta, A. (ed.). 2015. *Whales, Dolphins and Porpoises. A Natural History and Species Guide*. Chicago: University of Chicago Press.

Brakes, P. and M. P. Simmonds (eds). 2011. *Whales and Dolphins: Cognition, Culture, Conservation and Human Perceptions*. London and New York: Earthscan/Routledge and Taylor & Francis.

Burdin, A., O. Titova and E. Hoyt. 2014. *Humpback Whales of Russian Far East Seas. Photo-ID Catalog 2004–2014*. Moscow: Russian Geographical Society.

Carwardine, M. 1995, 2010. Whales, Dolphins and Porpoises. London: Dorling Kindersley.

Carwardine, M., E. Hoyt, P. Gill and E. Fordyce. 2006. *Whales, Dolphins & Porpoises*. Fog City Press, San Francisco.

Cell Press. 2015. The bowhead whale lives over 200 years: Can its genes tell us why? *ScienceDaily*, 5 January. sciencedaily.com/releases/2015/01/150105101421.htm

Evans, P.G.H. and J. A. Raga (eds.) 2001. Marine Mammals: Biology and Conservation. New York: Kluwer Academic / Plenum Publishers.

Fedutin, I.D., O. A. Filatova, E. G. Mamaev, A. M. Burdin and E. Hoyt. 2015. Occurrence and social structure of Baird's beaked whales, *Berardius bairdii*, in the Commander Islands, Russia. *Marine Mammal Science* 31(3): 853–865.

Filatova, O.A., I. D. Fedutin, T. V. Ivkovich, M. M. Nagailik, A. M. Burdin and E. Hoyt. 2009. The function of multi-pod aggregations of fish-eating killer whales (*Orcinus orca*) in Kamchatka, Far East Russia. *Journal of Ethology* 27(3):333-341.

Filatova, O.A., I. Fedutin, O. V. Titova, B. Siviour, A. M. Burdin and E. Hoyt. 2016. White killer whales (*Orcinus orca*) in the Western North Pacific. *Aquatic Mammals* 42(3):350-356.

Filatova, O. A., V. B. Deecke, J. K. B. Ford, C. O. Matkin, L. G. Barrett-Lennard, M. A. Guzeev, A. M. Burdin and E. Hoyt. 2012. Call diversity in the North Pacific killer whale populations: implications for dialect evolution and population history, *Animal Behaviour* 83, pp595-603.

Hoyt, E. 2011. *Marine Protected Areas for Whales, Dolphins and Porpoises: A World Handbook for Cetacean Habitat Conservation and Planning*. London and New York: Earthscan/Routledge and Taylor & Francis.

Hoyt, E. 2013. *Seasons of the Whale. Dark Shadows in the North Atlantic*. Nature Editions, North Berwick, U.K. pp.1-93.

Hoyt, E. 2014. *Creatures of the Deep. In Search of the Sea's Monsters and the World They Live In*. 2nd edition, rev. & expanded. Toronto: Firefly Books.

Hoyt, E. 2015. Whales through a new lens, *Hakai Magazine*, accessed Sept. 30, 2015. http://bit.ly/1Qx1v1O

Hoyt, E. and G. Notarbartolo di Sciara, eds. 2014. Report of the Workshop for the Development of Important Marine Mammal Area (IMMA) Criteria. Marseille, France, 22 Oct. 2013, IUCN Marine Mammal Protected Areas Task Force and International Committee on Marine Mammal Protected Areas, 20 pp.

Ivashchenko, Y. V. and P. J. Clapham. 2014. Too much is never enough: The cautionary tale of Soviet illegal whaling, *Marine Fisheries Review* 76:1-21.

Ivkovich, T., O. A. Filatova, A. M. Burdin, H. Sato and E. Hoyt. 2010. The social organization of resident-type killer whales (*Orcinus orca*) in Avacha Gulf, Northwest Pacific, as revealed through association patterns and acoustic similarity. *Mammalian Biology* 75:198-210.

Jefferson, T. A., R. Pitman, M. A. Weber, R. L. Pitman and U. Gorter. 2015. *Marine Mammals of the World. A comprehensive guide to their identification*, 2nd edition. London and San Diego: Academic Press.

Keane, M. et al. [29 co-authors] 2015. Insights into the evolution of longevity from the bowhead whale genome. *Cell Reports* 10, issue 1:112-122.

Kraus, S.D. and R. M. Rolland (eds.) 2010. *The Urban Whale. North Atlantic Right Whales at the Crossroads*. Cambridge: Harvard University Press.

Mann, J., R. C. Connor, P. L. Tyack and H. Whitehead. 1999. *Field Studies of Dolphins and Whales*. Chicago and London: University of Chicago Press.

McClain, C.R., M. A. Balk, M. C. Benfield, T. A. Branch, C. Chen, J. Cosgrove, A. D. M. Dove, L. C. Gaskins, R. R. Helm, F. G. Hochberg, F. B. Lee, A. Marshall, S. E. McMurray, C. Schanche, S. N. Stone and A. D. Thaler. 2015. Sizing ocean giants: patterns of intraspecific size variation in marine megafauna. PeerJ 3:e715 https://doi.org/10.7717/peerj.715

Morin, P. A., C. S. Baker, R. S. Brewer, A. M. Burdin, M. L. Dalebout, J. P. Dines, I. Fedutin, O. Filatova, E. Hoyt, J-L. Jung, M. Lauf, C. W. Potter, G. Richard, M. Ridgway, K. M. Robertson and P. R. Wade. 2016. Genetic structure of the beaked whale genus *Berardius* in the North Pacific, with genetic evidence for a new species. *Marine Mammal Science*. doi: 10.1111/mms.12345.

Perrin, W. F., B. Würsig and J. G. M. Thewissen (eds.). 2009. *Encyclopedia of Marine Mammals*, 2nd edition. San Diego: Academic Press.

Reeves, R. R., B. S. Stewart, P. J. Clapham, J. A. Powell and P. A. Folkens. 2008. *Guide to Marine Mammals of the World*. 2nd edition, New York: Knopf.

Rocha, Jr., R. C., P. J. Clapham and Y. V. Ivashchenko. 2015. Marine Fisheries Review, 76(4): 37–48.

Shirihai, H. and B. Jarrett. 2006. *Whales, Dolphins and Seals: A Field Guide to the Marine Mammals of the World*. London: A. & C. Black.

Simmonds, M. 2005. *Whales and Dolphins of the World*. London: New Holland.

Whitehead, H. 2003. *Sperm Whales: Social Evolution in the Ocean*. Chicago and London: The University of Chicago Press.

Whitehead, H. and L. Rendell. 2015. *The Cultural Lives of Whales and Dolphins*. Chicago and London: The University of Chicago Press.

Index

Note: **bold** page numbers indicate illustrations.